Springer Complexity

Springer Complexity is a publication program, cutting across all traditional disciplines of sciences as well as engineering, economics, medicine, psychology and computer sciences, which is aimed at researchers, students and practitioners working in the field of complex systems. Complex Systems are systems that comprise many interacting parts with the ability to generate a new quality of macroscopic collective behavior through self-organization, e.g., the spontaneous formation of temporal, spatial or functional structures. This recognition, that the collective behavior of the whole system cannot be simply inferred from the understanding of the behavior of the individual components, has led to various new concepts and sophisticated tools of complexity. The main concepts and tools – with sometimes overlapping contents and methodologies – are the theories of self-organization, complex systems, synergetics, dynamical systems, turbulence, catastrophes, instabilities, nonlinearity, stochastic processes, chaos, neural networks, cellular automata, adaptive systems, and genetic algorithms.

The topics treated within Springer Complexity are as diverse as lasers or fluids in physics, machine cutting phenomena of workpieces or electric circuits with feedback in engineering, growth of crystals or pattern formation in chemistry, morphogenesis in biology, brain function in neurology, behavior of stock exchange rates in economics, or the formation of public opinion in sociology. All these seemingly quite different kinds of structure formation have a number of important features and underlying structures in common. These deep structural similarities can be exploited to transfer analytical methods and understanding from one field to another. The Springer Complexity program therefore seeks to foster cross-fertilization between the disciplines and a dialogue between theoreticians and experimentalists for a deeper understanding of the general structure and behavior of complex systems.

The program consists of individual books, books series such as "Springer Series in Synergetics", "Institute of Nonlinear Science", "Physics of Neural Networks", and "Understanding Complex Systems", as well as various journals.

Understanding Complex Systems

Series Editor

J.A. Scott Kelso
Florida Atlantic University
Center for Complex Systems
Glades Road 777
Boca Raton, FL 33431-0991, USA

Understanding Complex Systems

Future scientific and technological developments in many fields will necessarily depend upon coming to grips with complex systems. Such systems are complex in both their composition (typically many different kinds of components interacting with each other and their environments on multiple levels) and in the rich diversity of behavior of which they are capable. The Springer Series in Understanding Complex Systems series (UCS) promotes new strategies and paradigms for understanding and realizing applications of complex systems research in a wide variety of fields and endeavors. UCS is explicitly transdisciplinary. It has three main goals: First, to elaborate the concepts, methods and tools of self-organizing dynamical systems at all levels of description and in all scientific fields, especially newly emerging areas within the Life, Social, Behavioral, Economic, Neuro- and Cognitive Sciences (and derivatives thereof); second, to encourage novel applications of these ideas in various fields of Engineering and Computation such as robotics, nano-technology and informatics; third, to provide a single forum within which commonalities and differences in the workings of complex systems may be discerned, hence leading to deeper insight and understanding. UCS will publish monographs and selected edited contributions from specialized conferences and workshops aimed at communicating new findings to a large multidisciplinary audience.

Alexander L. Fradkov

Cybernetical Physics

From Control of Chaos to Quantum Control

With 70 Figures

 Springer

Alexander L. Fradkov
Russian Academy of Sciences
Institute for Problems of Mechanical Engineering
Bolshoy Prospekt 61
199178 St. Petersburg, Russia
E-mail: alf@control.ipme.ru

Library of Congress Control Number: 2006933224

ISSN 1860-0832
ISBN-10 3-540-46275-9 Springer Berlin Heidelberg New York
ISBN-13 978-3-540-46275-0 Springer Berlin Heidelberg New York

This work is subject to copyright. All rights are reserved, whether the whole or part of the material is concerned, specifically the rights of translation, reprinting, reuse of illustrations, recitation, broadcasting, reproduction on microfilm or in any other way, and storage in data banks. Duplication of this publication or parts thereof is permitted only under the provisions of the German Copyright Law of September 9, 1965, in its current version. Violations are liable for prosecution under the German Copyright Law.

Springer is a part of Springer Science+Business Media
springer.com
© Springer-Verlag Berlin Heidelberg 2007

The use of general descriptive names, registered names, trademarks, etc. in this publication does not imply, even in the absence of a specific statement, that such names are exempt from the relevant protective laws and regulations and therefore free for general use.

Typesetting: by the author and techbooks using a Springer LaTeX macro package
Cover design: Erich Kirchner, Heidelberg

Printed on acid-free paper   SPIN: 11410027   54/techbooks   5 4 3 2 1 0

To my patient wife Olga

Preface

In this book a number of ideas and results related to the field of *cybernetical physics* – the scientific area aimed at the study of physical systems by cybernetical means – are presented. Although some publications in physical journals related to control problems appeared long ago, formation of a separate self-sustained area started as late as in the 1990s due to the explosion-like development of research in control of chaos, quantum control, and other areas. The number of publications has reached several thousands by the end of the century and it continues to grow very rapidly.

This book is, perhaps, the first attempt to present a unified exposition of the subject and methodology of cybernetical physics as well as solutions to some of its problems. A part of this book presents the limits of system transformation by means of control both for conservative and for dissipative systems based on Hamiltonian description of system dynamics.

A survey of various control applications in physics is given: control of chaos, controlled synchronization, control of spatiotemporal systems, control of molecular and quantum systems. An approach for building models of system dynamics based on control methods is discussed. The presented methods and results are illustrated by examples of new approaches to classical problems: Stephenson–Kapitsa pendulum, escape from a potential well, synchronization of coupled oscillators, control of chemical reaction with phase transition, controlled dissociation of molecules, controlled oscillations of complex crystalline lattices. Controlled pendulums appear in many parts of the book since pendulum models can be thought of as the "atoms of nonlinear physics".

The book is intended for a multidisciplinary audience: mostly only the understanding of basic concepts and results in linear algebra, calculus, and differential equations is needed. However, the style of exposition and the language used in cybernetics is somewhat more mathematically oriented than many physicists are accustomed to. Thus, a deeper understanding of the book may sometimes require an extra effort from the reader. A certain degree of familiarity with the basic concepts of control theory, dynamical systems theory, and information theory would facilitate reading. Concise and very readable exposition of those areas can be found in [58, 176, 177, 227, 419].

A difficulty of writing such books is the orientation toward a broader audience preventing a deepening of the exposition. Perhaps, physicists will not be satisfied with some places in the book, while control audience will not be satisfied with others. The author kindly asks readers for understanding and tolerance: the field is still very young, its common language is still under construction and solutions to many problems are yet to appear. Some open problems are mentioned in the text.

Although explanations of some basic concepts of control theory are given in the book, they cannot substitute for a good textbook for people with physical background which is yet to appear. References to the existing textbooks are included into the text.

It is the author's opinion that the area of cybernetical physics has been basically formed and it is the right time to get scientific communities, especially young scientists, acquainted with and interested in a new field. Such an opinion was encouraged by many of my colleagues during invited lectures to cybernetical and physical audiences in 1998–2004 in a number of universities and scientific centers, including the universities of Berlin, Bochum, California (San Diego), Duisburg, Eindhoven, Kumamoto, Kyoto, Linkoping, Melbourne, Moscow, Potsdam, Princeton, Seville, Southern California, St. Petersburg, Tokyo, Valencia; research institutes for Problems in Mechanics (Moscow), Control Sciences (Moscow), Santa Fe (USA), CESAME (Mexico), INRIA (France), RIKEN (Japan), SUPELEC (France); CNRS Center of Theoretical Physics in Marseille (France). I am grateful to colleagues and friends for invitation and encouragement.

The main content of the book is a result of author's research in the Laboratory "Control of Complex Systems" in the Institute for Problems of Mechanical Engineering of Russian Academy of Sciences (RAS) in St. Petersburg, supported by the RFBR grants 99-01-00672, 02-01-00765, 05-01-00869, by the programs for basic research of the Presidium of RAS # 19, # 22, NWO-RFBR program (project 047.011.2004.004) and by the program of support of young scientists and leading scientific schools (grant NSh-2257.2003.1).

The author takes a pleasant opportunity to thank colleagues and coworkers whose results got a reflection in the book: E.L. Aero, M.S. Ananyevskiy, B.R. Andrievsky, Yu.A. Astrov, I.I. Blekhman, P.Yu. Guzenko, A.A. Efimov, R.J. Evans, S.M. Khryaschev, A.M. Krivtsov, S.A. Kukushkin, A.V. Osipov, A.Yu. Pogromsky, V.V. Shiegin, A.S. Shiriaev. A sincere gratitude is expressed to an anonymous referee for numerous deep and helpful comments. Author is grateful to B.N. Wicks for his valuable help in proofreading and would be indebted to anyone with constructive comments or suggestions.

Saint Petersburg, *Alexander Fradkov*
May, 2006

Contents

1 Introduction: Physics and Cybernetics ... 1
   1.1 Looking into the past ... 1
   1.2 Control of chaos ... 2
   1.3 Control of molecules ... 4
   1.4 Physics and information ... 5
      1.4.1 Information is physical ... 5
      1.4.2 Physics from information ... 7
   1.5 Physics, animal, and machine ... 10
   1.6 Types of control ... 11
   1.7 What is the use of control in physics? ... 13
      1.7.1 Opinion of physicists ... 13
      1.7.2 Opinion of cyberneticians ... 15

2 Subject and Methodology of Cybernetical Physics ... 17
   2.1 Models of controlled systems ... 17
   2.2 Control goals ... 21
   2.3 Control algorithms ... 26
   2.4 Methodology ... 27
      2.4.1 Gradient method ... 28
      2.4.2 Speed-gradient method ... 29
      2.4.3 Feedback linearization ... 32
   2.5 Results: Laws of cybernetical physics ... 34

3 Control of Conservative Systems ... 37
   3.1 Control of energy for Hamiltonian systems ... 37
   3.2 Example: Controlled pendulum ... 41
   3.3 The swinging (small control) property ... 42
   3.4 Control of first integrals ... 43
   3.5 Control of generalized Hamiltonian systems ... 46

4 Control of Dissipative Systems ... 49
- 4.1 Excitability analysis of dissipative systems ... 49
  - 4.1.1 Excitability index ... 50
  - 4.1.2 Properties of excitability index ... 51
  - 4.1.3 Case of Euler–Lagrange systems ... 54
  - 4.1.4 Example: Excitation of the dumped pendulum ... 55
  - 4.1.5 Example: Excitation of the Duffing system ... 55
- 4.2 Feedback resonance ... 57
- 4.3 Excitability index of pendulum systems ... 60

5 Controlled Synchronization ... 67
- 5.1 Definitions of synchronization ... 67
  - 5.1.1 Evolution of the synchronization concept ... 68
  - 5.1.2 Synchronization of processes ... 69
  - 5.1.3 Examples ... 70
  - 5.1.4 Synchronization of systems ... 72
  - 5.1.5 Discussion ... 76
- 5.2 Controlled synchronization design ... 78
  - 5.2.1 Synchronization and convergence ... 79
  - 5.2.2 Synchronization and stabilization ... 80
  - 5.2.3 Synchronization and observers ... 82
  - 5.2.4 Synchronization of affine nonlinear systems ... 83
  - 5.2.5 Pecora–Carroll scheme ... 84
  - 5.2.6 Synchronization and speed-gradient ... 86
- 5.3 Adaptive synchronization ... 87
  - 5.3.1 Problem formulation ... 87
  - 5.3.2 Adaptive synchronization of two subsystems ... 87
  - 5.3.3 Conditions for control goal achievement ... 89
  - 5.3.4 Synchronization and adaptive observers ... 92
  - 5.3.5 Example: Information transmission using chaotic Chua system ... 95
- 5.4 Synchronization of two coupled pendulums ... 98

6 Control of Chaos ... 105
- 6.1 Introduction ... 105
- 6.2 Notion of chaos ... 106
  - 6.2.1 Definitions of chaos ... 106
  - 6.2.2 Criteria of chaos ... 109
  - 6.2.3 Delayed coordinates and Poincaré map ... 110
- 6.3 Models of controlled systems and control goals ... 111
- 6.4 Methods of controlling chaos: Continuous-time control ... 113
  - 6.4.1 Feedforward control by periodic signal ... 113
  - 6.4.2 Linearization of Poincaré map (OGY method) ... 116
  - 6.4.3 Delayed feedback (Pyragas method) ... 119
  - 6.4.4 Linear and nonlinear control ... 121

		6.4.5 Adaptive control 126

        6.4.5   Adaptive control 126
  6.5   Discrete-time control 127
  6.6   Generation of chaos (chaotization) 128
  6.7   Time and energy needed for control of chaos 129
  6.8   Applications in physics 133
        6.8.1   Control of turbulence 133
        6.8.2   Control of lasers 134
        6.8.3   Control of chaos in plasma 134
  6.9   Other problems 136

7 Control of Interconnected and Distributed Systems 137
  7.1   Models of controlled spatiotemporal systems 137
  7.2   Control of energy in sin-Gordon
and Frenkel–Kontorova models 139
  7.3   Control of wave motion in the chain of pendulums 142
        7.3.1   Modeling the chain of the pendulums 142
        7.3.2   Problem statement and control algorithm design 143
        7.3.3   Simulation results 144
  7.4   Control of oscillations in a complex crystalline lattice 148
        7.4.1   Modeling interaction of acoustic and optical modes 149
        7.4.2   Control law design 151
        7.4.3   Nonfeedback control 153
  7.5   Control of chaos in distributed systems 155
        7.5.1   Spatiotemporal systems 155
        7.5.2   Delayed systems 157
        7.5.3   Chaotic mixing 158

8 Control of Molecular and Quantum Systems 161
  8.1   Laser control of molecular dynamics 161
  8.2   Controlled dissociation of diatomic molecules (classical design) 164
        8.2.1   Control algorithm design 164
        8.2.2   Simulation results (classical model) 166
        8.2.3   Comparison of classical and quantum simulations 166
  8.3   Control of finite-level quantum systems 168

9 Control Algorithms and Dynamics of Physical Systems 173
  9.1   Integral and differential variational principles 173
  9.2   Examples of speed-gradient laws of dynamics 174
  9.3   Speed-gradient entropy maximization 176
  9.4   Onsager relations 179
  9.5   Discussion: Dynamics and the purpose 181

10 Examples 183
10.1 Controlled Stephenson–Kapitsa pendulum 183
10.2 Escape from a potential well and lossless communications..... 188
10.3 Feedback spectroscopy 192
10.4 Control of chemical reaction with phase transition 193
10.4.1 Problem statement 193
10.4.2 Adaptive control algorithm 194
10.4.3 Simulation results 196
10.5 Energy-like control of predator–prey system................. 199
10.6 Control of noise-induced transition......................... 202
10.6.1 The system model and problem statement 203
10.6.2 Control algorithm design 205
10.6.3 Control system analysis 206
10.6.4 Discussion.. 210

11 Conclusions: Looking into the Future 213

References.. 217

Index 239

1
Introduction: Physics and Cybernetics

1.1 Looking into the past

Encyclopedias define physics as the science studying Nature, specifically its most basic and universal properties. The age of physics transcends two millennia: the term "Physics," meaning "Nature" in Greek was introduced by Aristotle.

Cybernetics is much younger and the date of its birth is known precisely. Although the term was coined in Ancient Greece, the foundation of cybernetics as a science is associated with the publishing of Norbert Wiener's seminal book [446] in 1948. Wiener defined cybernetics as "the science of control and communication in the animal and the machine." Today cybernetics is understood as control theory in a broad sense, including different methods and approaches, such as identification, estimation, filtering, information theory, optimization, pattern recognition, etc. [432].

In the 20th century both physics and cybernetics experienced tremendous growth and contributed a lot into the development of modern science. However, cybernetical terms were rarely referred to in physical journals until recently and its influence on physical research has been negligible. The reason lies, perhaps, in the difference in methodologies of the two sciences. On the one hand, physics (e.g., mechanics) is a classical *descriptive* science. On the other hand, cybernetics (e.g., control theory) represents a paradigm for *prescriptive sciences* [89]. In other words, the main aim of physics is to describe and analyze a natural system or its behavior, while the aim of cybernetics is to find methods of transforming a system by means of controlling action in order to achieve its prescribed behavior. At first sight combining a phenomenon-oriented science with a methodology-oriented science seems hard if not impossible. Fortunately, developments of the last decade provide numerous evidences of the opposite. To present the most important of them will be the main task of this book.

It is worth noting that automatic and automated systems have been used in physical experiments for a long time and at present no serious physical experiment is performed without the use of automated equipment. However, an automatic system usually plays a secondary role, just providing a mean to achieve the desired mode of the experiment. Until recently no creative interaction of physics and control theory has been seen and no control theory methods have been directly used for discovering new physical effects and phenomena. Surprisingly, the situation changed in the 1990s when two new areas emerged: *control of chaos* and *quantum control*.

1.2 Control of chaos

A new avenue of research in physics began in the 1990s through results in control and synchronization of chaos. Chaotic system is a deterministic dynamical system exhibiting irregular, seemingly random behavior. Two trajectories of a chaotic system starting close to each other will diverge after some time (so-called "sensitive dependence on initial conditions").[1] Despite chaotic behavior seems unpredictable, it was found to be controllable. Edward Ott, Celso Grebogi, and James Yorke [331], from the University of Maryland discovered that even small feedback action can dramatically change behavior of a nonlinear system, e.g., turn chaotic motions into periodic ones and vice versa. The idea became popular in the physics community almost immediately and since 1990 hundreds of papers were published demonstrating the ability of small control, with or without feedback, to change dynamics of real or model systems significantly.

By 2003, the Ott, Grebogi, and Yorke's paper [331] has been quoted over 1300 times whilst the total number of the papers relating to control of chaos exceeded 4000 by the beginning of the 21st century. The number of papers published in peer-reviewed journals achieved 300–400 papers per year, see Fig. 1.2.1(a). The method proposed in [331] is now called the OGY-method after the authors' initials.

It is worth to note that in the 1980s the group of physicists from Moscow State University has demonstrated the possibility to transform a chaotic process into a periodic one by means of an external harmonic excitation [9, 10, 123, 249, 250]. For example, still in 1983 the possibility of transforming chaotic behavior of the Lorenz system into a periodic one under harmonic excitation was discovered [123]. Same property was discovered in [9, 10] for a model ecological system of fourth order by means of computer simulations. Nonetheless, the papers [9, 10, 123, 249, 250] did not trigger any stream of publications despite being translated and published in the West.

It is important that the results obtained were interpreted as discovering new properties of physical systems. Thousands of papers were published that examined and predicted properties of systems based on using control, identification, and other cybernetical methods. Notably, an overwhelming part of

---

[1] The definitions of chaos and main properties of chaotic systems will be given in Section 6.2.

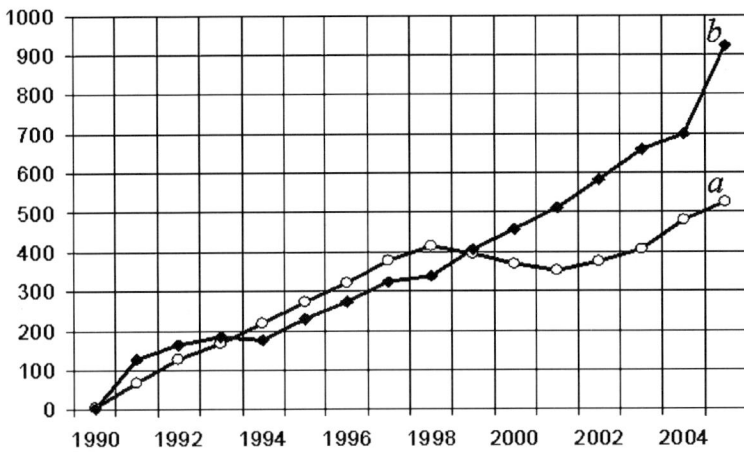

Fig. 1.2.1. Dynamics of publications in peer-reviewed journals (a) Control of chaos; (b) Quantum control (based on Science Citation Index data).

those papers were published in physical journals, its authors were representatives from the physical departments of the universities. This provides evidence for the existence of the new emerging field of research related to both physics and control, that of *Cybernetical Physics* [140, 141].

At that stage the potential of modern nonlinear control theory was still not realized, although the key role of nonlinearity was definitely appreciated. On the other hand, control community was not very active in discovering a new application area. The reason, perhaps was in that new problems often differed from conventional engineering control problems. Indeed, instead of a classical regulation problem (driving a controlled system trajectory to a desired point) or tracking problem (driving a controlled system trajectory to a desired motion) other control goals are of interest for physicists, such as creation of modes with partially specified properties (synchronization, transformation of a chaotic motion into a periodic one, etc.) At the same time, more strict restrictions are imposed on controlling action which correspond to the physically motivated requirement of minimum intervention in the natural evolution of the physical system ("small control" requirement).

It has become clear that such type of control goals are important not only for control of chaos, but also for control of a broader class of oscillatory processes. This lead to development of a unified control framework for oscillatory (including chaotic) systems [25, 164]. The next step was to pose a general control problem of studying properties of physical systems that can be created or modified by means of (small) feedback actions [139–141, 144, 148].

Recently, an interest in the application of cybernetical methods to the search of new physical effects has been observed in other fields of physics and mechanics, such as control of quantum systems, control of lasers, control of plasma, vibration control, control of particle beams, control in thermody-

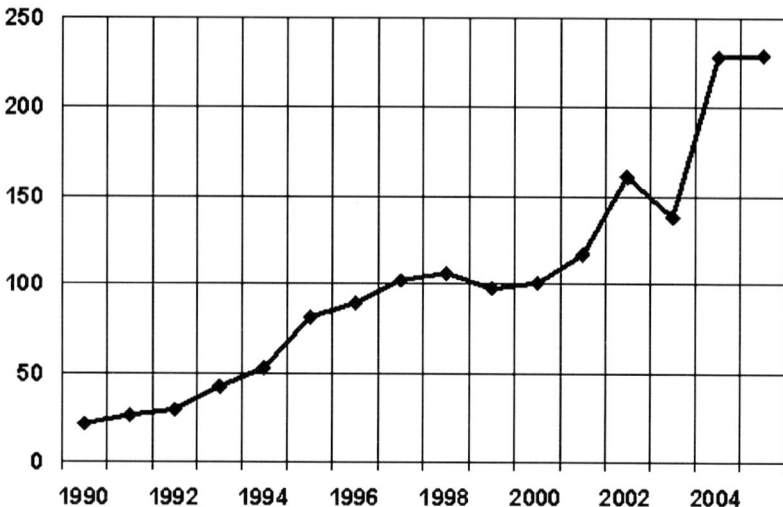

Fig. 1.2.2. Dynamics of publications in the journals of American Physical Society (*Physical Review A, B, C, D, E*; *Physical Review Letters*), having the term "control" in the title of the paper.

namics, etc. As a consequence, a number of control-related papers in physical journals is growing rather rapidly, see Fig. 1.2.2. Especially fast growth during last decade is observed in control of molecular and quantum systems, see Fig. 1.2.1(*b*).

1.3 Control of molecules

It is conceivable that molecular physics was the area where ideas of control appeared first. One may trace its roots back to the Middle Ages, where alchemists were seeking ways to change a natural course of chemical reactions whilst attempting to transform lead into gold. The next milestone was put by the famous British physicist James Clerk Maxwell. In 1871, he introduced a hypothetical being with ability to measure velocities of gas molecules and to direct fast molecules to one part of the vessel, keeping slow molecules in the other part. It produces a temperature difference between the two parts of the vessel which seems to contradict the Second Law of thermodynamics. Now this being is known under the name "Maxwell's Demon" introduced by another famous physicist William Thomson (Lord Kelvin). The apparent breaking of the Second Law helps to its better understanding. Now, after more than a century of fruitful life, Demon is even more active than in the past. In the 20th century Demon was exploited by L. Szillard, D. Gabor, L. Brillouin, and others who studied the interplay between energy and information [88, 263, 355, 423]. It has helped humanity to realize that any measurement or computation requires some dissipation of energy. Attempts to diminish dissipation led to the idea of quantum computers [122, 234, 314]. In recent papers the issues of experimental implementation of the Maxwell's Demon are discussed, particularly at the quantum–mechanical level [52, 275].

In the end of the 1980s–beginning of the 1990s rapid development of laser industry led to appearance of ultrafast, the so-called *femtosecond* lasers. A new generation of lasers have the ability to generaste pulses with duration of about a few femtoseconds and even less (1 fs = 10^{-15} s). The duration of such a pulse is comparable with the period of a molecule's natural oscillation. Therefore, femtosecond laser can be, in principle, used as a means, for controlling single molecules and atoms. A consequence of such an application is a possibility of realizing an alchemist's dream to change the natural course of chemical reactions. Besides, control is an important part of many recent nanoscale applications: nanomotors, nanowires, nanochips, nanorobots, etc.

Using the apparatus of modern control theory, new horizons in studying interaction of atoms and molecules may open new ways and possible limits for intervention into intimate processes of the microworld may be discovered.

1.4 Physics and information

1.4.1 Information is physical

The exchange of matter and exchange of energy are the two types of interaction between systems and their environment commonly studied in physics. In the second half of the 20th century a lot of interest was attracted by the third type of interaction, namely, exchange of information. N. Wiener considered study of information and communication as a part of his cybernetics [446]. Systematic development of the information theory started with the seminal work of Claude Shannon [401]. However, importance of the notion of information was recognized well before the beginning of the computer revolution [196, 321].

First attempts to understand the relations between information and other characteristics of physical systems started when attempting to elucidate Maxwell's Demon [423]. It turned out that the Maxwell's Demon has become a helpful being to clarify interrelation of information with other physical quantities. It was Leo Szillard [423] who first established connection between entropy and information by means of some quantitative measure of information, closely related to what was later proposed by Shannon. Obviously, decrease of the system entropy due to the ordering (control) of molecules requires some measurements, i.e., extraction of information about state of molecules. Szillard showed that in order to receive information one needs to increase entropy and its increase cannot be less than its decrease caused by the ordering of molecules. Later, this conclusion was extended to more general physical processes and finally a universal conservation law for the sum of entropy and information was established [88, 220, 236, 355, 356].

A history of relations between information and energy is somewhat dramatic. An importance of understanding interrelations between energy exchange and information transmission was recognized still in the 1940s. It was Shannon who derived in 1948 that at least $kT \ln 2$ units of energy is needed to transmit a unit (1 bit) of information in a linear bosonic channel with additive noise [401], where T is absolute temperature and k is the Boltzmann constant. This is just an energy required to make a signal distinguishable above the thermal background. In 1949 John von Neumann extended that statement to the following principle: *any computing device, natural or artificial must dissipate at least $kT \ln 2$ energy per elementary transmission of 1 unit of information* [313]. Next contribution into linking energy and information was made by Rolf Landauer: " Data processing operation has irreducible thermodynamic cost if and only if it is logically irreversible" [257]. Thus, existence of reversible communications and computation would apparently contradict the von Neumann's principle. However, it was not clear *how* to transmit information with minimal energy.

A clarification was made again by Landauer who considered the case of nonlinear channels in which information is carried in the internal state of a material body (e.g., a bistable molecule with two states separated by a high-energy barrier). He established both for classical [258] and for quantum [260] case that a bit can be transmitted without minimal unavoidable cost. It means that a communication network with only a small loss of transmitted information may require only a small amount of energy for information transmission. However, the problem of organizing an efficient transmission is still to be solved, since the solution proposed by Landauer resembles an existence theorem and does not suggest a way to design such a transmitter.

For example, how to put a carrier of a bit of information into one or another well of a bistable potential? It is a control problem which is not easy to solve on a molecular level. Despite existence of publications arguing against Landauer's principle (see, e.g., [319]) it is clear that there exist physical limits for information transmission and sophisticated methods are needed to approach them. Some hints how to solve the problem can be made based on the energy control methods described below, see Section 10.2.

As an outcome of discussions an importance of another principle advocated by Landauer has been widely recognized: information should be treated as a physical quantity, like energy or entropy [259, 261]. Quoting Landauer, "Information is tied to a physical representation and therefore to restrictions and possibilities related to laws of physics." By the end of the 20th century a new area of physics has been shaped named *information physics*. A number of surveys and monographs has been published, treating information as a measure of interaction between physical systems [176, 221, 263, 356]. Special attention is focused on the issues of quantum information and quantum computation [122, 234, 314].

1.4.2 Physics from information

Thermodynamics is another field of physics where methods of information theory and control theory are actively applied for several decades. Even its classical results can be interpreted in "cybernetical spirit." Let us trace the brief history of the thermodynamic ideas from the cybernetic viewpoint, following [215].

The basics of thermodynamics were stated by Sadi Carnot in 1824. He considered a heat engine which operates by drawing heat q_1 from a source which is at thermal equilibrium at temperature T_{hot}, and delivering useful work W. Carnot saw that, in order to operate continuously, the engine requires also a cold reservoir with the temperature $T_{\text{cold}} < T_{\text{hot}}$ to which some heat q_2 can be discharged. By simple logic he established the famous

Carnot's Principle. No heat engine can be more efficient than a reversible one operating between the same temperatures.

It implies that maximal efficiency of any heat engine depends only on the temperatures $T_{\text{cold}} < T_{\text{hot}}$. In fact, it was nothing but the solution to an optimal control problem: maximum work can be extracted by a reversible machine and the value of extracted work depends only on the temperatures of the source and the bath. Later, Kelvin introduced his absolute temperature scale (Kelvin scale) and accomplished the next step, evaluating the Carnot's reversible efficiency

$$\eta_{\text{Carnot}} = 1 - \frac{T_{\text{cold}}}{T_{\text{hot}}}. \tag{1.4.1}$$

Being motivated with the fact that any irreversible process has smaller efficiency, Rudolph Clausius introduced a new function of the system state – *entropy* – in 1865 and interpreted the Carnot's principle as follows: in the change from one thermal equilibrium state to another, the total entropy of all bodies involved cannot decrease:

$$S_{\text{fin}} \geq S_{\text{ini}}. \tag{1.4.2}$$

Such a formulation is known as the fundamental Clausius' statement of the Second Law of thermodynamics. Later, Josiah Willard Gibbs and Maxwell extended the scope of applicability as well as the meaning of the Second Law (1.4.2). It has become a fundamental physical law establishing the general direction in which an irreversible process will go. Instead of Clausius' weak statement that the total entropy of all bodies involved "tends" to increase, Gibbs made the strong prediction that it will increase, up to the maximum value permitted by whatever constraints (conservation of energy, volume, mole numbers, etc.) are imposed by the experimental arrangement and the known laws of physics [215]. Note that this prediction was formulated by Gibbs only implicitly. Its explicit formulation first appeared in 1957 in the famous paper by Edwin T. Jaynes [214] under the name of

Maximum Entropy Principle (MEP). A system evolves to a state with maximum value of entropy compatible with all imposed constraints.

8   1 Introduction: Physics and Cybernetics

The procedure of MEP usage is simple and basically consists in application of the Lagrange multipliers method. MEP has obviously a cybernetic spirit, since it relates the model of physical system with information available to the researcher. In the following years MEP was triumphantly applied in different areas of natural science, social science and engineering [193, 225, 289, 449]. It was used for building and analysis of mathematical models for numerous systems of different origin outside thermodynamics. It was applied to examination of self-organization and complexity phenomena [194]. However, most works were devoted to study of stationary systems over infinite time interval, while for practical purposes it is important to know possibilities and limitations of the system evolution for finite times as well as under other types of constraints caused by a finite amount of resources available.

The pioneer works devoted to evaluation of finite time limitations for heat engines were published by I. Novikov in 1957 [320] and F.L. Curzon and B. Ahlborn in 1975 [113]. It was shown independently in [113, 320] that the efficiency at maximum power per cycle of a heat engine coupled to its surroundings through a constant heat conductor is

$$\eta_{\text{MaxP}} = 1 - \sqrt{\frac{T_{\text{cold}}}{T_{\text{hot}}}}. \qquad (1.4.3)$$

Relation between (1.4.3) and classical Carnot efficiency is seen from Fig. 1.4.3.

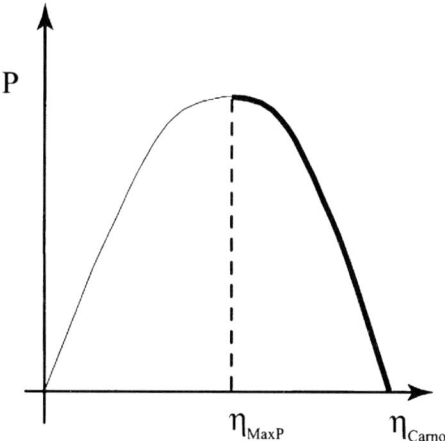

Fig. 1.4.3. Extracted power as a function of efficiency for a simple Novikov–Curzon–Ahlborn-type heat engine.

Note that the Novikov–Curzon–Ahlborn process is also optimal in the sense of minimal dissipation. Otherwise, if the dissipation degree is given, the process corresponds to the maximum entropy principle. Later, the results of [113, 320] were extended and generalized for other criteria and for more complex situations based on the modern optimal control theory. As a result the whole direction in thermodynamics arose known under the names *optimization thermodynamics, finite-time thermodynamics*, or *control thermodynamics*, see monographs and surveys [17, 22, 59, 67, 297, 387, 410]. Control thermodynamics is certainly a part of cybernetical physics. Recently a move in the opposite direction has become noticeable: thermodynamics is used for design of new control algorithms [13, 195, 454]. Mutual influence of control and thermodynamics is growing rapidly.

The maximum entropy principle has been generalized to take into account a prior information about the probability distribution of the system. Its generalized version employs the concept of *relative entropy* (or *cross-entropy* or *Kullback–Leibler divergence*) [248]. The divergence between discrete distributions $\rho(x)$ and $\bar{\rho}(x)$ is defined as follows:

$$K(\rho, \bar{\rho}) = \sum_{i=1}^{N} \rho(x_i) \ln \frac{\rho(x_i)}{\bar{\rho}(x_i)}. \qquad (1.4.4)$$

It is easy to see that $K(\rho, \bar{\rho}) \geq 0$ and $K(\rho, \bar{\rho}) = 0$ if and only if $\rho = \bar{\rho}$. Let $\bar{\rho}$ be the distribution characterizing the initial knowledge about the system state. Then a reasonable way to predict the real distribution of the state is to choose ρ providing minimal value of $K(\rho, \bar{\rho})$ consistent with other physical constraints, e.g., with conservation laws. If *a priori* information about the state is absent, then the plausible choice of $\bar{\rho}$ is the uniform distribution $\bar{\rho}(x_i) = 1/N$. In this case, obviously, $I(\rho, \bar{\rho}) = -S(\rho) + \ln(N)$ and minimization of the divergence (1.4.4) yields the same result as maximization of the entropy of the distribution ρ, i.e., standard MEP is recovered.

Interesting versions of MEP-like variational principles are the principle of *Minimum Fisher Information* (MFI) and the principle of *Extreme of Physical Information* (EPI) proposed recently by B. Roy Frieden [167–169]. They have a more broad field of applicability than MEP and are sometimes advocated as unifying principles of physics.

The Fisher information for continuous distribution with smooth density $\rho(x)$ is defined as follows

$$F(\rho) = \int [\rho(x)']^2 \frac{dx}{\rho(x)} = \int [\ln \rho(x)']^2 \rho(x) dx.$$

The two important quantities are closely related [439]: for $\bar{\rho}(x) = \rho(x + \Delta x)$ the cross-entropy $2K(\rho', \bar{\rho})/(\Delta x)^2$ is an approximation of $F(\rho)$ for $\Delta x \to 0$. Therefore, in many cases extremization of the cross-entropy is equivalent to extremization of the Fisher information.

The history of aforementioned principles shows another way for penetration of cybernetical ideas into physics. These ideas, first born in thermodynamics due to intimate relations of the concepts of entropy and information are now expanding far beyond thermodynamics and influence the very foundations of physics. The slogan "It from bit" proclaimed in 1989 by famous

American physicist, inventor of "black holes" John Archibald Wheeler [445] attracts more and more followers who are eager to derive physical laws from information-related postulates. Being not able to resist this temptation we suggest an alternative path in the same direction. In Chapter 9, the derivation of some physics-related results from the control-related speed-gradient method of Section 2.4.2 will be described following [135, 147].

1.5 Physics, animal, and machine

Application of cybernetic methods in biological physics is very important and deserves a special discussion. A plenty of articles dedicated to this topic can be found in physical journals. Cybernetic methods are also applied in related areas: neuroscience, medicine, etc. A substantial part of research deals with analysis of biological and biomedical time series. A number of methods of spectral analysis, smoothing and filtering, pattern recognition developed in cybernetics literature are used for analysis of electrocardiograms (ECG), electroencephalograms (EEG), and other biorhythm records. Methods of identification and parameter estimation are often used for building mathematical models of organisms and their parts.

N. Wiener started his research in this area a few years before he proclaimed the birth of cybernetics. With his colleagues he studied analogies between behavior of living beings and engineering systems [373]. His findings provided a basis for the understanding of new science. Later it was reflected in the title of his book: "Cybernetics or control and communication in the animal and the machine." Such analogies proved to be useful for extracting common features of signal processing, coordination and control in biological and engineering systems. In the 1960s a new scientific area emerged: *biological and medical cybernetics*. The biomedical cybernetics is still an area of active research. Its results can be found in the journals and conference proceedings entitled: "Biological Cybernetics," "Medical Cybernetics," "Neuroscience," and others.

Methods of cybernetics and nonlinear dynamics are recently intensively applied for the investigation of complex behavior in biological systems [307, 396]. It is shown by many authors that such phenomena as synchronization and chaos play important role in signal processing and regulation in neural system and brain activity [2, 350].

A large body of research is devoted to natural or artificial neural networks. Analysis and control of chaotic dynamics of neural networks are of considerable interest. Neural nets in chaotic modes are used to model information storage and pattern recognition functions of the brain [237, 309, 424]. Such networks may consist of not only artificial neurons, but also of other nonlinear systems with controllable chaotic behavior, e.g., natural neurons or chemical oscillators, see [198]. A fundamental question concerns mechanisms which explain how a population of neurons, whose individual activity is chaotic and uncorrelated can form functional circuits with regular and stable behavior. This has been addressed both theoretically and experimentally in [2, 364, 365].

Among other interesting problems, synchronization of respiratory and cardiac rhythms is worth mentioning [350, 389]. Studying synchronization and coordination of muscular activity is very important for control of motion and posture of animals and humans [60, 61, 291, 317, 374].

Joint functioning of mechanism and organism, of natural and artificial, of a robot and a human being is the problem of utmost importance for future. We know about artificial muscles, artificial human arms and legs. Implanted pacemakers and microchips for stimulating and monitoring of the human body subsystems are becoming a part of the medician's toolbox. All mentioned examples are nothing but preliminary examples of *cyborgs* – cybernetical organisms, born at the pages of science fiction books.

The aforementioned researches have a number of practical applications. One of the most exciting and promising of early applications of the control of chaos was treating cardiac arrhythmia see [121, 173, 429]. Development of a smart feedback pacemaker seemed to create a novel approach in cardiology. Another prospective area is study and treating of epilepsy by the methods of chaos control [391]. It is based upon the well-known correlation between epileptic seizure and appearance of extra synchronies between neuronal spike sequences. Concepts of phase synchronization was used for analysis of electroencephalograms (EEG) [306].

Application of cybernetic methods in biophysics and life sciences is an area of rapidly growing activity. However, we do not have time and place enough to discuss this issue further.

1.6 Types of control

Certainly, some studies based on cybernetic methods were going on in physics well before the 1990s. For example, investigation of parameter-dependent models in nonlinear dynamics demands for examination of bifurcation modes (bifurcation means qualitative change in system behavior with quantitative change of its parameters). It looks like the input parameter (denote it by u) becomes a variable rather than a constant quantity: $u = u(t)$. The terminology reflecting this fact suggests to call such a new input a *control parameter* or *control variable* (The term *control variable* would fit better since the quantity $u = u(t)$ may vary in time).

Actually, creating or eliminating a bifurcation can be interpreted as a lower, trivial form of control. However, in this case the control variable is constant in time: $u(t) = $ const, see Fig. 1.6.4(a). Solutions to this kind of problems usually provide information about possible limits of control, i.e., shows possible behaviors of the system for different parameter values.

Optimization problems where the goal is to find the value of control (input) parameter providing a minimum or maximum value for the given system

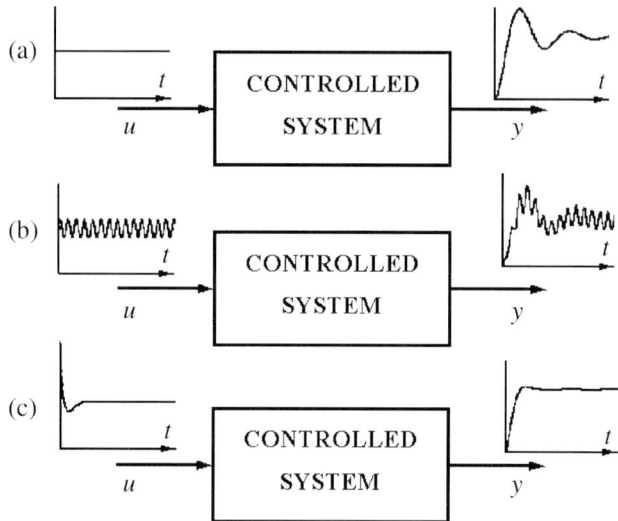

Fig. 1.6.4. Types of control. (a) Constant (trivial) control; (b) feedforward control; (c) feedback control.

performance index also fall into this class. Again, admissible control functions should be constant in time: $u(t) = $ const.

Another class of physical problems is the study of a system under action of input variables (disturbances) depending only on time: $u = u(t)$, see Fig. 1.6.4(b). Problems of such kind include spectroscopic studies, vibration analysis in mechanics and acoustics, some parts of oscillation and wave theories where inputs are harmonic functions: $u(t) = A\sin(\omega t)$. In the second half of the 20th century a new branch of mechanics – *vibrational mechanics*, studying behavior and properties of mechanical systems and materials under fast oscillating action was developed [73]. An advantage of such kind of control is that one does not need any measurements of the system state or observables. It is most important for fast processes at microscopic, e.g., molecular level where any measurement is hard if not impossible to perform.

The areas where similar problems are intensively studied are control theory and control engineering. For example, frequency response of a linear dynamical system is used by engineers to analyze controlled systems since it reflects behavior of a system under harmonic input signal or force. In order to evaluate frequency response experimentally, harmonic signals with different frequencies are applied to a system and its output oscillation amplitude and phase shift with respect to the input are measured. The theory of *vibrational control* studies how to change a system behavior by means of applying fast oscillating signal or force [62, 292]. A control action not using feedback (nonfeedback control) is usually called *open loop* or *feedforward* control. It is interesting to note that the principle of phase stability operating in accelerators, proposed independently by Veksler and McMillan in 1944–1945, can be seen as control by means of an oscillating action with a slowly varying frequency. Such a method of control (sometimes called the dynamic autoresonance [130, 311] or chirping [274]) makes it possible to accumulate the energy of a nonlinear system while the system remains in the resonant mode, and at present is used to excite atoms, plasma, molecular systems, etc.

However, the choices of control action in form $u = $ const or $u = u(t)$ by no means cover all possible ways of controlling a system. Moreover, there exists much more powerful way, namely, *feedback* control. Feedback control is based on using measurements of the system output $y = y(t)$, see Fig. 1.6.4(c). It can be expressed in the form $u(t) = U(y(t))$ or $u(t) = U(y(t), t)$. It is not an exaggeration to say that outstanding achievements of control theory during last half of century have become possible owing to development of efficient feedback design methods.

Note that *internal feedback* can be found in many physical systems and plays an essential role in modeling of a system. However, feedback as a mean of investigation of a system requires an *external feedback*. Such usages of feedback have drawn the attention of physicists just recently. Perhaps, it was the area of controlling chaos where feedback was first systematically used for studying possibilities of changing behavior of the systems. Unlike conventional usage of feedback for engineering applications, where most efficient way to achieve the *desired behavior* is sought, a physical study is aimed at evaluation and explanation of a class of possible *behaviors* of a physical system achievable by applying a feedback.

Feedback control is a very powerful means of changing a systems behavior. It may even cause an antipathy to feedback-based study: one may think that the object of study is changed if feedback is applied. To avoid confusions, serious restrictions should be imposed on controlling action, e.g., a *small control* requirement. On the other hand, apparently any experiment with a physical system leads to some change in it, no matter if a feedback is used or not. For example, it is well known in quantum mechanics that an observation may even destroy a system. Appealing to physical intuition should help to find proper interpretation of experimental data avoiding erroneous inference and discovering artefacts.

One of the main goals of this book is to examine possible ways of using feedback for exploration of physical system properties. In other words, an alternative subtitle of the book could be: "What can be done by feedback?"

1.7 What is the use of control in physics?

1.7.1 Opinion of physicists

It is interesting to know what is the opinion of physicists about a new area and what specific features and advantages they find. It is easy to extract such an information scanning the control-related papers in physical journals. Let us quote a few papers to provide a sample of a few opinions.

> The aim of the researches is twofold: – to create a particular product that is unattainable by conventional chemical means; – to achieve a better understanding of atoms and molecules and their interactions. (Rabitz H. et al. Whither the Future of Controlling Quantum Phenomena? *Science*, 2000, 288, 824–828.)

> There are two fields of application of controlling friction. Obviously there will be technological applications for reducing vibration and wear. But controlling friction experiments can also be used to increase our understanding of the physics of dry friction. For example, using these methods one can measure the effective friction force as a function of the sliding. (Elmer F.J. Controlling friction. *Physical Review E*, V. 57, 1998, R490–R4906.)

> We have summarized some recently proposed applications of control methods to problems of mixing and coherence in chaotic dynamical systems. This is an important problem both for its own intrinsic interest and also from the point of view of applications. Those methods provide insights also into the origin of mixing and unmixing behavior in natural systems. (Sharma A., Gupte, N. Control methods for problems of mixing and coherence in chaotic maps and flows. *Pramana – J. of Physics*, V. 48, 1997, 231–248.)

> We develop novel diagnostics tools for plasma turbulence based on feedback. This ... allows qualitative and quantitative inference about the dynamical model of the plasma turbulence. (Sen A.K. Control and diagnostic uses of feedback. *Physics of Plasmas*, V. 7, 2000, 1759–1766.)

> In a world dominated by electronics, the ability not only to remove chaos where it is not wanted, but also to make more flexible circuitry by exploring chaos and its control could have a tremendous impact on our lives (Ditto W.L., Spano M.L., Lindner J.F. Techniques for control of chaos. *Physica D*, V. 86, 1995, 198–211.)

> We believe that controlled stochastic resonance may be useful in applications as diverse as the cancellation of power-line frequencies in very sensitive magnetic sensing, applications with super-conducting quantum interference devices and vibration control in nonlinear mechanical devices, as well as in the context of electromagnetic field interactions with neuronal tissue, where control of internal thresholds is possible and the selective suppression of specific frequencies could potentially be beneficial. (Ločher M., et al. Theory of controlling stochastic resonance. *Phys. Review E*, V. 62, 2000 (1), 317–327.)

1.7 What is the use of control in physics?

We have analyzed synchronization in a system describing acoustic gravity waves which arises in atmospheric physics. By using the mechanism of adaptive controller which requires minimum knowledge about the structure of the system, we have found that the system synchronizes with its error converging to zero. Applications to Lorenz system and Chua circuits have been done and it is expected that such analysis will be helpful in revealing the relation between two different non-linear processes in plasma or two similar non-linear processes under varied spatio-temporal situation. (S. Banerjee, P. Saha, A.R. Chowdhury. On the application of adaptive control and phase synchronization in non-linear fluid dynamics. *International Journal of Non-Linear Mechanics*. V. 39, 2004, 25–31)

1.7.2 Opinion of cyberneticians

The opinion of physicists have much in common with the views of control theorists and engineers working in the same area.

Cybernetic concepts describe physico-chemical, biological, and social phenomena with equal success. (V. Turchin. *The Phenomenon of Science. A cybernetic approach to human evolution.* Columbia University Press, 1977)

The fact that control and dynamical systems communities had fundamentally different aims led to some misunderstanding, in which control theorists saw the dynamical systems work as naive and dynamical systems workers thought that control theorists only knew how to control using elephants instead of butterflies. (G. Chen, X. Dong, From Chaos to Order, *World Scientific*, 1998)

The control of chaos is one of the most popular topics in the complexity research. Physicists play major role in this area. Motivated by deeper study of chaos control, they learned seriously nonlinear control theory. They appreciated its potential and try to expand the boundary of physics and control beyond the chaos control. (H. Kimura. A Personal View of Control in Future. *ACPA Newsletter* No. 7. Oct 1999)

Quantum Control is emerging as a challenging discipline, with applications ranging from quantum computation, to metrology and spectroscopy. Its typical tasks, state steering and quantum operation realization, have been effectively reformulated in terms of classical control problems. Thus, not only physicists but an increasing number of control engineers start to work in the field, applying their well established methods to these newborn problems. (F. Ticozzi, A. Ferrante. Linear algebraic techniques for quantum dynamical decoupling. *Proc. IEEE Conf. Decision and Control - Europ. Control Conf.* Seville, 2005.)

The above quotations give a nice idea of what is going on. Both respectable scientific communities – physicists and cyberneticians are excited about the prospects of a new field. New challenges and new horizons make a new field equally attractive for people coming from different areas.

At the same time the cooperation of experts in different areas face a number of challenges, the main being mutual understanding of people with quite a diverse scientific backgrounds. To work out a common language in the area some basic problems should be considered from a unified viewpoint. An attempt to do so is undertaken in the next chapters for a class of energy control and related problems.

2
Subject and Methodology of Cybernetical Physics

Now we are in position to define *Cybernetical Physics* as the branch of science studying physical systems by cybernetical means [140, 141, 147, 148]. In this chapter its subject, particularly control problems for physical systems will be discussed. In order to characterize control problems related to cybernetical physics, the main classes of controlled plant models, control objectives (goals), and admissible control algorithms will be specified. In order to outline the methodology of cybernetical physics, some typical methods used for solving the problems and typical results in the field will be described.

The term "cybernetical" rather than "cybernetic" is suggested by analogy with other fields in the crosses of sciences: mathematical physics, chemical physics, etc. However, an aposcopic (truncated) version "cybernetic" is also acceptable since it is already in use in engineering with other nouns: cybernetic methods, cybernetic systems, etc.

2.1 Models of controlled systems

A formal statement of any control problem begins with a model of the system to be controlled (plant) and a model of the control objective (goal). Even if the plant model is not given (like in many real world applications) it should be determined in some way. The system models used in cybernetics are similar to traditional models of physics and mechanics with one difference; the inputs and outputs of the model should be explicitly specified. The following main classes of models are considered in the literature related to control of physical systems.

The most common class consists of continuous systems with lumped parameters described in state space by differential equations

$$\dot{x} = F(x, u), \qquad (2.1.1)$$

where x is n-dimensional vector of the state variables[1]; $\dot{x} = d/dt$ stands for the time derivative of x; u is m-dimensional vector of inputs (control variables). Components of the state vector x are denoted as $x_1, x_2, \ldots, x_n$, while the components of controlling vector u are denoted as $u_1, u_2, \ldots, u_m$. Therefore, (2.1.1) is a compact notation for the system of ordinary differential equations

$$\frac{dx_i}{dt} = F_i(x_1, x_2, \ldots, x_n, u_1, u_2, \ldots, u_m), \quad i = 1, 2, \ldots, m. \tag{2.1.2}$$

Vector–function $F(x, u)$ is usually assumed continuously differentiable to guarantee existence and uniqueness of the solutions of (2.1.1) at least at some time interval close to the initial point $t = 0$. However, a time interval where the model (2.1.1) is considered is usually not predefined and some additional requirements may be needed to guarantee that solutions of (2.1.1) are well defined for all $t \geq 0$.

It is important to note that model (2.1.1) encompasses two physically different cases:

A. *Coordinate (force) control.* The input variables represent some physical variables (forces, torques, intensity of electrical or magnetic fields, etc.) For example a model of a controlled oscillator (pendulum) can be put into the form

$$J\ddot{\varphi} + r\dot{\varphi} + mgl \sin \varphi = u, \tag{2.1.3}$$

where $\varphi = \varphi(t)$ is the angle of deflection from vertical; J, m, l, r are physical parameters of the pendulum (inertia moment $J = ml^2/2$, mass, length, friction coefficient); g is gravity acceleration; $u = u(t)$ is the controlling torque. The description (2.1.3) is transformable into the form (2.1.1) with the state vector $x = (\varphi, \dot{\varphi})^T$ as follows:

$$\begin{cases} \dot{x}_1 = x_2 \\ \dot{x}_2 = -rJ^{-1}x_2 - mgl/J \sin x_1 + 1/Ju(t). \end{cases} \tag{2.1.4}$$

B. *Parametric control.* The input variables represent change of physical parameters of the system, i.e., $u(t) = p - p_0$, where p_0 is the nominal value of the physical parameter p. For example, let the pendulum be controlled by changing its length: $l(t) = l_0 + u(t)$. If $l(t)$ is a slowly varying variable then the model, instead of (2.1.3), becomes

---

[1] Hereafter the following notations are used: $\mathbb{R}^n$, $\mathbb{C}^n$ are real and complex n-dimensional vector spaces, respectively; $x \in \mathbb{R}^n$ is a real n-dimensional vector (column); $x = \text{col}(x_1, x_2, \ldots, x_n)$ stands for a column vector with the components $x_1, x_2, \ldots, x_n$; the Euclidean norm of the vector $x \in \mathbb{R}^n$ is denoted as $|x| = (x_1^2 + x_2^2 + \ldots + x_n^2)^{1/2}$; if X is the vector (matrix), then X^T stands for the transposed vector (matrix). Particularly, if X is the column vector, then X^T is the row vector. Notation I_n stands for $n \times n$ unity matrix; $\square$ marks the end of definition, example, or remark; ■ marks the end of the proof.

2.1 Models of controlled systems

$$J\ddot{\varphi} + r\dot{\varphi} + m(l_0 + u(t))\sin\varphi = 0. \qquad (2.1.5)$$

If the rate of the length change $\dot{l}(t)$ cannot be neglected, it is natural to choose it as a new controlling variable $v(t)$:

$$\dot{l}(t) = v(t). \qquad (2.1.6)$$

In this case the dynamics model derived from Euler–Lagrange equation takes, instead of (2.1.5), the form

$$m(l(t))^2\ddot{\varphi} + 2ml(t)u(t)\varphi + r\dot{\varphi} + mgl(t)\sin\varphi = 0 \qquad (2.1.7)$$

and the plant model is described by equations (2.1.6), (2.1.7).

Although in some papers the difference between the cases A and B is emphasized, for the purpose of studying a nonlinear system (2.1.1) the difference is not very important.[2]

If external disturbances are present, we need to consider more general time-varying models

$$\dot{x} = F(x, u, t). \qquad (2.1.8)$$

On the other hand, many nonlinear control problems can be described using more simple affine in control models:

$$\dot{x} = f(x) + g(x)u. \qquad (2.1.9)$$

The model should also include the description of measurements, i.e., the l-dimensional vector of output variables (observables) y should be defined, for example, as a function of the system state:

$$y = h(x). \qquad (2.1.10)$$

If the outputs are not defined explicitly, it will be assumed that all the state variables are available for measurement, i.e., $y = x$. Using notation $y = h(x, u)$ means that some input variables are also available for measurement.

An important example of output for physical systems is *energy*. For example, the energy of the pendulum (2.1.3) is defined as follows: $H = 0.5J(\dot{\varphi})^2 + mgl(1 - \cos\varphi)$. Therefore it is not sufficient to consider only linear functions $h(x)$ as it is accustomed in conventional control theory.

Note that using notation (2.1.10) implicitly implies that the measurement does not influence the dynamics of the physical system, or that such an influence is negligible. Of course, such an assumption does not hold for many

---

[2] It makes sense to treat differently the case of coordinate control and the case of parametric control only for linear systems. The reason is that a linear system with a linear coordinate feedback remains linear, while a linear system with linear parametric feedback leaves the class of linear systems and the system becomes bilinear. However the class of nonlinear systems (2.1.1) is closed with respect to all nonlinear feedbacks, and both coordinate and parametric controls can be treated in a similar way.

physical processes, especially for quantum processes where macroscopic measurement may even destroy a microscopic system. All such problems need individual consideration.

Note also that the notion of the state in control theory differs from the notion of the state used in thermodynamics. Hereafter we understand the state as a set of variables such that their dynamics are described by a system of first order differential equations. All other observables are assumed to be expressed as functions of the state variables. Among examples are mechanical systems where state can be defined as the set of coordinates and velocities of all particles, or quantum–mechanical system where the state may be defined as the wave function obey the first order Shrödinger equation in the Hilbert space. In some problems the description of the controlled system requires differential equations on the manifolds. However, such problems will not be addressed in this book.

For many systems discrete-time state–space models are used:

$$x_{k+1} = F_d(x_k, u_k), \qquad (2.1.11)$$

where $x_k \in \mathbb{R}^n, u_k \in \mathbb{R}^m, y_k \in \mathbb{R}^l$, are state, input, and output vectors at kth stage of the process. Then the model will be defined by the mapping F_d. Using a discrete-time model may be convenient even if the process $x(t)$ is a continuous-time one, but the measurements are taken at discrete-time instants (sampling times) t_k, $k = 1, 2, \ldots$. Then $x_k = x(t_k)$, $u_k = u(t_k)$, $y_k = y(t_k)$.

The natural correspondence between the continuous-time and discrete-time systems may be established if some time instant t_k is specified for each kth step of the discrete-time system. It induces the correspondence $x_k = x(t_k), u_k = u(t_k)$. The connection between the right-hand side of the discrete-time and continuous-time models is not uniquely defined in general and may be specified by additional conventions [177]. The most common way is to assume that $t_k = k\Delta t$, where $\Delta t > 0$ is the discretization step (sampling interval) and to define the input $u(t)$ of the continuous-time system as follows: $u(t) = u_k$ for $t_k \leq t < t_{k+1}$. It means that the input to the continuous-time system is piecewise constant which is usually the case for computer-controlled systems having sampler as discrete-to-analog (D-A) converter. However the exact expression for $x(t_k)$ is still not available since it requires integration of the differential system over the sampling interval. The simplest approximate solution is given by the Euler numerical integration formula:

$$x(t_{k+1}) \approx x(t_k) + \Delta t F(x(t_k), u(t_k), t_k).$$

Setting $F_d(x, u, k) = x + \Delta t F(x, u, k\Delta t)$, $h_d(x, u, k) = h(x, u, k\Delta t)$ we obtain the discrete-time model with a local error of the order Δt^2. The error over the time interval $[0, T]$ can be evaluated as $e^{LT}\Delta t$, where L is the Lipschitz constant of $F(x, u, t)$ in x along the solution $x(t), 0 \leq t \leq T$.

Note that for the linear systems (2.1.1) is possible to achieve the exact discretization using the Cauchy formula for exact solution of the linear differential equations with piecewise constant inputs:

$$F_d(x,u) = Px + Qu, \quad h_d(x,u) = h(x,u), \qquad (2.1.12)$$

where $P = \exp(A\Delta t), Q = A^{-1}(P - I_n)B$. However, it involves computation of the matrix exponential which in general can only be approximated.

In Chapter 7 control problems for distributed (spatiotemporal) systems described by partial differential equations or their discrete analogs will be examined. Finally, we will need delay–differential models

$$\dot{x} = F(x(t), x(t-\tau), u(t), u(t-\tau_u)), \qquad (2.1.13)$$

and delay–difference models

$$x_{k+1} = F_d(x_k, x_{k-1}, ..., x_{k-\tau}, u_k, ..., u_{k-\tau_u}). \qquad (2.1.14)$$

To determine solutions of system (2.1.13) on some time interval $[t_0, t_1)$ it is necessary to specify the initial state function $\overline{X}_0 = \{x(s), t_0 - \tau \le s \le t_0\}$ in addition to the input function $\overline{U}_0 = \{u(s), t_0 - \tau \le s \le t_0\}$. Delay may appear in the system model for different reasons. Typically, it may be caused by spatially extended system location (transport delay) or by artificial delay introduced by control (delayed feedback, see Section 6.4.3.)

In what follows we will assume that all the models under consideration satisfy conditions guaranteeing existence of their solutions at least at some interval starting from given initial conditions for all $t \ge t_0$. For simplicity we will also assume that $t_0 = 0$ whenever possible.

2.2 Control goals

It is natural to classify control problems by their control goals. We list here five kind of goals.

A. *Regulation.* Regulation (often called stabilization or positioning) is the most common and simple control goal. Regulation is understood as driving the state vector $x(t)$ or the output vector $y(t)$ to some equilibrium state x_* (respectively, y_*). Due to presence of various uncertainties, it is convenient to eliminate time from formulation of the control goal and to consider an idealized control goal as the limit relation

$$\lim_{t \to \infty} x(t) = x_* \qquad (2.2.15)$$

or limit relation

$$\lim_{t \to \infty} y(t) = y_*. \qquad (2.2.16)$$

In the presence of additive disturbances achievement of the control goals (2.2.15) and (2.2.16) is impossible in general and one should replace them by the limit relations for the upper limit (maximum limit over all subsequences) of the error:

$$\overline{\lim_{t\to\infty}} |x(t) - x_*| \le \Delta \qquad (2.2.17)$$

or

$$\overline{\lim_{t\to\infty}} |y(t) - y_*| \le \Delta, \qquad (2.2.18)$$

where Δ is the maximum value of admissible error. If the controlled system is under action of stochastic disturbances (noise), it is reasonable to introduce the averaged goal

$$\overline{\lim_{t\to\infty}} \mathbf{M}|x(t) - x_*| \le \Delta \qquad (2.2.19)$$

or

$$\overline{\lim_{t\to\infty}} \mathbf{M}|y(t) - y_*| \le \Delta, \qquad (2.2.20)$$

where $\mathbf{M}$ (mean) is the symbol of mathematical expectation (averaging).

The goals (2.2.15)–(2.2.20) are harder to achieve if the desired equilibrium state x_* is unstable in the absence of control action. Such a case is typical for control of chaotic systems. It is also possible that in the absence of control action the goal state x_* is not an equilibrium. However it does not introduce extra complication; it just means that control action may not disappear when the trajectory approaches x_*.

Note that it is not necessary to use Euclidean norm in the formulations (2.2.15)–(2.2.20). Other norms in the vector space $\mathbb{R}^n$ are also possible, e.g., weighting different coordinates of x or y differently. More generally, any nonnegative function $Q(x,t) \ge 0$ can be used in the formulations (2.2.15)–(2.2.20) instead of a norm:

$$\overline{\lim_{t\to\infty}} Q(x(t), t) \le \Delta, \qquad (2.2.21)$$

$$\overline{\lim_{t\to\infty}} \mathbf{M} Q(x(t), t) \le \Delta. \qquad (2.2.22)$$

Formulations (2.2.21), (2.2.22) allow to express goals related to convergence of only a part of the state coordinates, so-called partial stability, or stability with respect to a function [157, 437].

B. *Tracking.* State tracking is driving a solution $x(t)$ of (2.1.1) to the prespecified function of time $x_*(t)$, i.e., fulfillment of the relation

$$\lim_{t\to\infty} [x(t) - x_*(t)] = 0 \qquad (2.2.23)$$

for any solution $x(t)$ of (2.1.1) with initial conditions $x(0) = x_0 \in \Omega$, where Ω is given set of initial conditions. Similarly, output tracking is driving the output $y(t)$ to the desired output function $y_*(t)$, i.e.,

$$\lim_{t\to\infty} [y(t) - y_*(t)] = 0. \qquad (2.2.24)$$

The desired output function $y_*(t)$ may be interpreted as the *command* or *reference signal*. It may be either given explicitly before the system starts functioning or it may be measured online by some measurement device. Alternatively, $y_*(t)$ may depend on the motion of some auxiliary system called *reference model* or *model of the goal*. In the latter case the problem of finding a controller ensuring the goal (2.2.23) or (2.2.24) is referred to as *model reference control problem*.

For example, a typical problem of chaos control can be formulated as tracking of an unstable periodic solution (orbit). In this case $x_*(t)$ is the T-periodic solution of the free (uncontrolled, i.e., $u(t) = 0$) system (2.1.1) with initial condition $x_*(0) = x_*^0$, i.e., $x_*(t+T) = x_*(t)$ for all $t \geq 0$.

The key feature of the control problems for physical systems is that the goal should be achieved by means of sufficiently small control. A limit case is stabilization of a system by an arbitrarily small control. Solvability of this task is not obvious if the trajectory $x_*(t)$ is unstable, like for the case of chaotic systems, see [331].

A special case of the above problems is stabilization of the unstable equilibrium x_*^0 of system (2.1.1) with $u = 0$, i.e., stabilization of x_*^0, satisfying $F(x_*^0, 0) = 0$. Again, it is similar to a usual regulation problem with an additional restriction that we seek for "small control" solutions. However, such a restriction makes the problem far from standard: even for a simple pendulum. Nonlocal solutions to the stabilization problem for the upright equilibrium by means of small control were obtained just recently, see [408]. The class of admissible control laws can be extended by introducing dynamic feedback described by differential or time-delayed models. Similar formulations hold for discrete and time-delayed systems.

C. *Generation (excitation) of oscillations.* The third class of control goals corresponds to the problems of *excitation* or *generation* of oscillations. Here, it is assumed that the system is initially at rest. The problem is to find out if it is possible to drive it into an oscillatory mode with the desired characteristics (energy, frequency, etc.) In this case the goal trajectory of the state vector $x_*(t)$ is not prespecified. Moreover, the goal trajectory may be unknown, or may even be irrelevant to the achievement of the control goal. Such problems are well known in electrical, radio engineering, acoustics, laser, and vibrational technologies – wherever it is necessary to create an oscillatory mode for the system. Such a class of control goals can be related to problems of dissociation, ionization of molecular systems, escape from a potential well, chaotization, and other problems related to growth of the system energy and its possible phase transition. Sometimes such problems can be reduced to tracking, but the reference trajectory $x_*(t)$ in these cases are not necessarily periodic and may be unstable.

To formalize the excitation of oscillation problems it is convenient to introduce a scalar goal function $G(x)$ and specify the goal as achieving the limit equality
$$\lim_{t \to \infty} G(x(t)) = G_* \qquad (2.2.25)$$
or inequality for the lower limit of the goal function value

$$\underline{\lim}_{t\to\infty} G(x(t)) \geq G_*. \qquad (2.2.26)$$

In many cases the total energy of mechanical or electrical oscillations can serve as the goal function $G(x)$.

D. *Synchronization.* The fourth important class of control goals corresponds to *synchronization* (more accurately, *controlled synchronization* as distinct from *autosynchronization* or *self-synchronization*). Generally speaking, synchronization is understood as concurrent change of the states of two or more systems or, perhaps, concurrent change of some quantities related to the systems, e.g., equalizing of oscillation frequencies [73–75]. If the required relation is established only asymptotically, one speaks about *asymptotic synchronization*. If synchronization does not exist in the system without control (for $u = 0$) we may pose the problem as finding the control function which ensures synchronization in the closed-loop system, i.e., synchronization may be a control goal. Synchronization problem differs from the model reference control problem in that some phase shifts between the processes are allowed that are either constant or tend to constant values. Besides, in a number of synchronization problems the links between the systems to be synchronized are bidirectional ones. In such cases the limit mode (synchronous mode) in the overall system is not known in advance. A detailed discussion of synchronization problem statements will be presented in Chapter 5.

A simple way to formulate the control goal, corresponding, e.g., to asymptotic synchronization of the two system states x_1 and x_2 is to express it as the limit relation:

$$\lim_{t\to\infty} [x_1(t) - x_2(t)] = 0. \qquad (2.2.27)$$

In the extended state space $x = \{x_1, x_2\}$ of the overall system, relation (2.2.27) implies convergence of the solution $x(t)$ to the diagonal set $\{x : x_1 = x_2\}$. Asymptotic identity of the outputs or, more generally, of the values of some quantity $G(x)$ for two systems can be formulated as follows

$$\lim_{t\to\infty} [G(x_1(t)) - G(x_2(t))] = 0. \qquad (2.2.28)$$

Often it is convenient to rewrite the goals (2.2.23), (2.2.24), (2.2.25), (2.2.27), or (2.2.28) in terms of appropriate goal function $Q(x,t)$ as follows:

$$\lim_{t\to\infty} Q(x(t), t) = 0. \qquad (2.2.29)$$

For example, to reduce goal (2.2.27) to the form (2.2.29) one may choose $Q(x) = |x_1 - x_2|^2$. Instead of Euclidean norm other quadratic functions can also be used. In the case of the goal (2.2.23) the goal function $Q(x,t) = [x - x_*(t)]^T \Gamma [x - x_*(t)]$, where Γ is positive definite symmetric matrix can be used. The freedom of choice of the goal function can be utilized for design purposes.

E. *Modification of the limit sets (attractors) of the systems.* The last class of the control goals is related to modification of some quantitative characteristics of the limit behavior of the system. It includes such specific goals as
   – changing the type of the equilibrium (e.g., transformation of an unstable equilibrium into a stable one or vice versa);
   – changing the type of the limit set (e.g., transformation of a limit cycle into a chaotic attractor or vice versa, changing fractal dimension of the limit set, etc.);
   – changing the position or the type of the bifurcation point in the parameter space of the system.

Investigation of the above problems started in the end of the 1980s with the works on bifurcation control [3, 442] and continued in the works on control of chaos. Ott, Grebogi, and Yorke [331] and their followers introduced a new class of control goals, not requiring any quantitative characteristic of the desired motion. Instead, the desired qualitative type of the limit set (attractor) was specified, e.g., control should provide the system with a chaotic attractor.[3]

Development of new approaches to such problems is stimulated by new applications to laser and chemical technologies, in telecommunications, in biology and medicine. For example, functioning of the laser after its transition into a chaotic (multimode) regime can be restored by means of introducing a small optical feedback. This leads to increase of the power of coherent radiation. On the contrary, in chemical technologies the chaoticity property is important for good mixing in chemical reactors, leading to faster reaction and better quality of products. In this case a reasonable goal is to increase the degree of chaoticity. An example from medical research is applying control of chaos to treatment of some type of arrhythmias using feedback pacemakers [87, 173]. Irregularity degree of the cardiac rhythm is adjusted by means of applying stimulating pulses in appropriate time instants. In this case the control goal is maintaining the desired degree of irregularity. The goal functions expressing the desired chaoticity degree can be constructed based on standard chaoticity measures like Lyapunov exponents, fractal dimensions, entropy, etc.

In addition to the main control goal some additional goals or constraints may be specified. A typical example is the "small control" requirements: control function should have small power or should provide small expenditure of energy. Such a restriction is needed to avoid "violence" and preserve inherent properties of the system under control. This is important to ensure elimination of artefacts and adequate study of the system.

Mathematically the small control requirement may be expressed as restriction $\|u(\cdot)\| < \Delta$, where $\|u(\cdot)\|$ is some norm of the control function, while $\Delta > 0$ is a prespecified value (threshold).

---
[3] Definitions of chaos, attractor, and main properties of chaotic systems will be introduced in Section 6.2.

The possibility to achieve the goal may depend also on initial conditions. If the goal is achieved for any initial conditions, one may speak about global achievability of the goal. Otherwise, a set of admissible initial conditions Ω should be given, so that the goal is achieved for any solution $x(t)$ of the system (2.1.1) with admissible controls and initial conditions from Ω, i.e., for $x(0) \in \Omega$. Note that standard control goals in conventional control theory are regulation and tracking. The rest of the aforementioned goals are not typical for conventional control theory because they do not completely specify the desired behavior of the system. These classes of control problems belong to the area of the so-called *partial control* which has rapidly developed recently [157, 437]. It is important that the above goals should be achieved without significant intervention to the system dynamics, i.e., the design of the control algorithms should meet the *small control* or *weak control* requirement.

Let us now discuss some methods of control problems solution.

2.3 Control algorithms

Through the perusal of physical journals, many articles may be found using the terms "control," "controllable" in the following intuitive sense. Let examination of a system be made via variation of one parameter, called control, input or bifurcation parameter. Let examination of the system be performed for different values of input parameter which itself is constant as a function of time. Changes of such a parameter create changes to some characteristics of the system behavior, sometimes called output parameter. The system is called controllable if the range of changes of an output parameter under admissible changes of input parameters covers the values, corresponding the desired regimes of the system behavior.

Strictly speaking, "control" in the above-mentioned sense can hardly be called "control" in conventional control theory. It is just a possibility of control, a preliminary study of the system aimed at right control problem statement. Such a study may results in finding the value of the input parameter corresponding to the desired value of output. In reality however, even if the value of the input parameter is properly evaluated, keeping such a constant parameter value may be ineffective for achievement of the goal for a dynamical system.

For example, consider again the problem of stabilizing the unstable equilibrium $\varphi = \pi$ of the pendulum (2.1.3). Let the control action be constant: $u(t) \equiv 0$. Equilibrium condition for the point $\varphi = \pi$ implies $u(t) = 0$. However, because of instability of the equilibrium $\varphi = \pi$ any deflection of initial conditions or any external disturbance leads to violation of the control goal, no matter how small is a disturbance.

To find a more efficient way of control, one may try a time-dependent (time-varying) control action. To be precise, if a control parameter is time-varying, it cannot be called a parameter, it should be called a variable. Any rule allowing to calculate the value of the control action $u(t)$ in each time instant t is usually called a *control algorithm*. A number of control algorithms that can be found in the literature provide control variable depending solely on time: $u = u(t)$. Such kind of control action is called *program* or *command* control action, and the rule itself belongs to the class of so-called *open-loop* or *feedforward* control algorithms. The feedforward control may also depend on initial state of the system:

$$u(t) = U(t, x_0), \qquad (2.3.30)$$

where $x_0 = x(0)$.

Still more possibilities are provided with control algorithms using results of the system state or output measurements. Such an algorithm may have the form of *state feedback*

$$u(t) = U(x(t)) \qquad (2.3.31)$$

or the form of *output feedback*

$$u(t) = U(y(t)). \qquad (2.3.32)$$

All three types of control (constant, feedforward, and feedback) may be helpful in physical problems. Implementation of a feedback control requires additional measurement devices working in real time which are often hard to install. Therefore, studying the system may start with application of inferior forms of control: time-constant and then feedforward control. The possibilities of changing the system behavior by means of feedback control can then be studied.

2.4 Methodology

The methodology of cybernetical physics is based on control theory. Typically, some parameters of physical systems are unknown and some variables are not available for measurement. From the control viewpoint this means that control design should be performed under significant uncertainty, i.e., methods of robust or adaptive control should be used. A variety of design methods have been developed by control theorists and control engineers for both linear and nonlinear systems [157, 245, 464]. Methods of partial control, control by weak signals, etc. have also been developed [164, 437]. Below a few fairly general yet simple approaches to control of complex nonlinear systems are described briefly. Firstly, the gradient method applicable to discrete-time control is presented. Secondly, the speed-gradient method applicable to continuous-time systems is described. Thirdly, the feedback linearization method allowing in some cases to reduce nonlinear problems to linear ones is outlined. All three approaches can be used for nonlinear, robust, and adaptive control. They will be used in consequent chapters of this book.

2.4.1 Gradient method

Numerous systems in physics, biology, economics, and other areas can be described by discrete-time dynamical models. Even if a natural system is functioning in continuous time, its variables are often available for measurement or control only at some discrete sampling time instants and therefore, system model can be presented in a discrete-time form. Consider a class of controlled systems described by the discrete-time state–space model:

$$x_{k+1} = F(x_k, u_k), \qquad y_k = h(x_k, u_k), \qquad (2.4.33)$$

where $x_k \in \mathbb{R}^n$ is the value of the state vector at the kth step of system functioning, $y_k \in \mathbb{R}^l$ is the corresponding value of the output, $u_k \in \mathbb{R}^m$ is the kth value of the input (control) action. The vector functions F and h are assumed to be well defined for all values of states and inputs. In the case when the model (2.4.33) describes behavior of a continuous-time system measured at some sampling instants $k = 0, 1, 2, \ldots$, the variables can be interpreted as follows: $x_k \in \mathbb{R}^n$ is the value of the state vector $x(t)$ at the sampling instant t_k; $y_k \in \mathbb{R}^l$ is the value of the output measured at the sampling instant t_k, $u_k \in \mathbb{R}^m$ is the value of the input (control) applied to the system at the sampling interval $t_k \leq t < t_{k+1}$, $k = 0, 1, 2, \ldots$.

Let the goal function $Q(x) \geq 0$ be given and the control goal be specified as

$$Q(x_{k+1}) \leq \Delta, \quad \text{when} \quad k > k_* \qquad (2.4.34)$$

where $\Delta > 0$ is the prespecified threshold value. The gradient method of control algorithm design consists of two stages. At the first stage, the reduced goal function depending on the number of the step is calculated, substituting (2.4.33) into (2.4.34):

$$Q_k(u) = Q(F_k(x_k, u)) \qquad (2.4.35)$$

The reduced goal function directly depends on u. At the second stage the gradient vector

$$\nabla_u Q_k(u) = \text{col}\left(\frac{\partial Q_k(u)}{\partial u^{(1)}}, \ldots, \frac{\partial Q_k(u)}{\partial u^{(m)}}\right)$$

is calculated and the control algorithm

$$u_{k+1} = u_k - \gamma_k \nabla_u Q_k(u_k), \qquad (2.4.36)$$

where $\gamma_k \geq 0$, is the algorithm parameter (step size) is derived.

The algorithm (2.4.36) makes the current control correction $\Delta u_k = u_{k+1} - u_k$ along the descent direction of the current goal function $Q_k(u)$. The idea of the gradient method comes from optimization theory. However, in optimization problems the objective function does not depend on k. It is worth noting that there is no reason to use more complicated algorithms for control of a dynamical system because at every step the goal function may change.

A simple algorithm does not necessarily have simple applicability conditions. To formulate such conditions we use the so-called method of the *recursive goal inequalities* proposed by V.A. Yakubovich in 1966 [85, 132]. The key point of the method is to introduce a deadzone into the algorithm, i.e., to choose $\gamma_k = 0$ if the goal inequality (2.4.34) is fulfilled. The precise formulation of the applicability conditions can be found in [85, 164]. Essentially, three main conditions should be fulfilled: (A) the function $Q_k(u)$ is convex in u; (B) there exists a common solution $u = u_*$ to the system of the goal inequalities $Q_k(u) < \Delta$, $k = 0, 1, 2, \ldots$; (C) the choice of the gain γ_k takes into account the deadzone: if the current inequality $Q_k(u_k) \leq \Delta$ holds, then $\gamma_k = 0$ is chosen.

Note that it often happens that the right-hand side of the algorithm (2.4.36) depends on the whole nonmeasurable state vector x_k. There are two standard ways to treat such problems. The first is to include an additional dynamical system (so called *observer*), which performs an online estimation of the unknown state vector. The second is to replace the state–space model (2.1.11) of the controlled system by the input–output model:

$$y_{k+1} = \Phi(y_k, \ldots, y_{k-n}, u_k, \ldots, u_{k-n+1}). \tag{2.4.37}$$

Then at the kth step one will need to evaluate control in the form $u_k = U(y_k, \ldots, y_{k-n}, u_{k-1}, \ldots, u_{k-n+1})$ which is easier to design.

2.4.2 Speed-gradient method

A continuous-time counterpart of the gradient method is the so called *speed-gradient (SG) method*. Like the gradient method for discrete-time systems, SG-method is intended for control problems where control goal is specified by means of a goal function.

Consider a nonlinear time-varying system

$$\dot{x} = F(x, u, t) \tag{2.4.38}$$

and control goal

$$\lim_{t \to \infty} Q(x(t), t) = 0, \tag{2.4.39}$$

where $Q(x,t) \geq 0$ is a smooth goal function.

In order to design control algorithm the scalar function $\dot{Q} = \omega(x, u, t)$ is calculated, that is, the speed (rate) of changing $Q_t = Q(x(t), t)$ along trajectories of (2.4.38): $\omega(x, u, t) = \partial Q(x, t)/\partial t + [\nabla_x Q(x, t)]^T F(x, u, t)$. Then it is needed to evaluate the gradient of $\omega(x, u, t)$ with respect to input variables: $\nabla_u \omega(x, u, t) = (\partial \omega / \partial u)^T = (\partial F / \partial u)^T \nabla_x Q(x, t)$. Finally, the algorithm of changing $u(t)$ is determined according to the differential equation

$$\frac{du}{dt} = -\Gamma \nabla_u \omega(x, u, t), \qquad (2.4.40)$$

where $\Gamma = \Gamma^{\mathrm{T}} > 0$ is a positive definite gain matrix, e.g., $\Gamma = \mathrm{diag}\{\gamma_1, \ldots, \gamma_m\}$, $\gamma_i > 0$. The algorithm (2.4.40) is called *speed-gradient (SG) algorithm*, since it suggests to change $u(t)$ proportionally to the gradient of the speed of changing Q_t.

The origin of the algorithm (2.4.40) can be explained as follows. In order to achieve the control goal (2.4.39) it is desirable to change $u(t)$ in the direction where $Q(x(t), t)$ decrease. However, it may be a problem since $Q(x(t), t)$ does not depend on $u(t)$ directly. Instead one may try to decrease $\dot{Q}$, in order to achieve the inequality $\dot{Q} < 0$, which implies decrease of $Q(x(t), t)$. The speed $\dot{Q} = \omega(x, u, t)$ generically depends on u explicitly which allows to write down (2.4.40). The speed-gradient algorithm can be also interpreted as a continuous-time counterpart of the gradient algorithm, since for small sampling step size the direction of the gradient is close to the direction of the speed-gradient.

Let us illustrate speed-gradient design methodology for a class of tracking control problems for controlled systems linear in the inputs:

$$\dot{x} = A(x, t) + B(x, t)u, \qquad (2.4.41)$$

where $x(t) \in \mathbb{R}^n$ is the state vector, $u(t) \in \mathbb{R}^m$ is vector of controlling variables (inputs) which may be either physical quantities or adjustable parameters, $A(x, t)$ is n-vector, $B(x, t)$ is $n \times m$-matrix. Let the control goal have the form

$$\lim_{t \to \infty} [y(t) - y_*(t)] = 0, \qquad (2.4.42)$$

where $y(t) = h(x(t)) \in \mathbb{R}^l$ is l-vector of regulated variables (outputs), $y_*(t) \in \mathbb{R}^l$ is the goal trajectory (desired trajectory) of the outputs. It is clear that the goal (2.4.42) has equivalent form (2.4.39) if the goal function $Q(x, t)$ is chosen as follows:

$$Q(x, t) = \frac{1}{2}[y - y_*(t)]^{\mathrm{T}} P[y - y_*(t)], \qquad (2.4.43)$$

where P is symmetric positive-definite $l \times l$-matrix.

For the purpose of control algorithm design rewrite equation (2.4.41) in the form

$$\dot{x} = A(x, t) + \sum_{i=1}^{m} B_i(x, t) u_i, \qquad (2.4.44)$$

where u_i are components of the vector $u \in \mathbb{R}^m$ and $B_i(x, t) \in \mathbb{R}^n$ are columns of the matrix $B(x, t)$. Then the rate (speed) of changing $Q(x(t), t)$ along trajectories of the system (for constant u) is as follows:

$$\omega(x, u, t) = [y - y_*(t)]^{\mathrm{T}} P[CA(x, t) + CB(x, t)u - \dot{y}_*(t)], \qquad (2.4.45)$$

where $C = C(x,t) = \partial G(x,t)/\partial x$. Taking the gradient of (2.4.45) in u we obtain the speed-gradient and the speed-gradient algorithm in the following form

$$\nabla_u \omega(x,u,t) = B(x,t)^\mathrm{T} C^\mathrm{T} P[y - y_*(t)], \tag{2.4.46}$$

$$\frac{du}{dt} = -\Gamma B(x,t)^\mathrm{T} C^\mathrm{T} P[y - y_*(t)]. \tag{2.4.47}$$

To simplify design, the gain matrix Γ is often chosen as diagonal matrix ($\Gamma = \mathrm{diag}\{\gamma_i\}$) or scalar matrix ($\Gamma = \gamma I$) where ($\gamma_i, \gamma$) are positive numbers. For special case of the system linear in inputs the algorithm (2.4.47) is nothing but the classical *integral control law*.

In a similar way the so-called speed-gradient algorithm in finite form is designed

$$u(t) = u_0 - \Gamma \nabla_u \omega(x(t), u(t), t), \tag{2.4.48}$$

where u_0 is some initial value of control variable, e.g., $u_0 = 0$). Algorithm (2.4.48) is a generalization of classical *proportional control law*.

More general form of speed-gradient algorithms is sometimes useful:

$$u(t) = u_0 - \gamma \psi(x(t), u(t), t), \tag{2.4.49}$$

where $\gamma > 0$ is the scalar gain parameter and vector-function $\psi(x,u,t)$ satisfies the so-called *pseudogradient condition*

$$\psi(x,u,t)^\mathrm{T} \nabla_u \omega(x,u,t) \geq 0 \tag{2.4.50}$$

for all x, u, t. Special case of (2.4.49) is called *sign-like* or *relay-like* algorithm:

$$u(t) = u_0 - \gamma \,\mathrm{sign}\, \nabla_u \omega(x(t), u(t), t), \tag{2.4.51}$$

where sign of a vector is understood component-wise: for a vector $z = \mathrm{col}\,(z_1, \ldots, z_m)$ sign z is defined as sign $z = \mathrm{col}\,(\mathrm{sign}\, z_1, \ldots, \mathrm{sign}\, z_m)$.

In order to make a reasonable choice of the control algorithm parameters the applicability conditions should be verified. The main conditions are: convexity of the function $\omega(x,u,t)$ in u and existence of "ideal" control u_* such that $\omega(x, u_*, t) \leq 0$ for all x (attainability condition). More precise formulations and mathematical proofs can be found in [29, 134, 135, 157, 164].

The speed-gradient algorithms can be modified to take into account constraints. For example, let the equality constraint be given

$$g(x(t), u(t), t) = 0, \tag{2.4.52}$$

where g is a smooth scalar function, and a scalar control function $u(t)$ is to be chosen such that (2.4.52) is satisfied for all $t \geq 0$. The modified (constrained) SG-algorithm in differential form is as follows:

$$\dot{u}(t) = -\gamma \nabla_u \omega(x(t), u(t), t) - \lambda(t) \nabla_u g(x(t), u(t), t), \tag{2.4.53}$$

where the Lagrange multiplier $\lambda(t)$ is chosen to satisfy condition $\dot{g} = 0$, that is,

$$\lambda(t) = \frac{-\gamma \nabla_u \omega(x(t), u(t), t) + \nabla_x g^{\mathrm{T}} F(x(t), u(t), t) + \partial g/\partial t}{|\nabla_u g(x(t), u(t), t)|^2}. \qquad (2.4.54)$$

Initial condition $u(0)$ should satisfy constraint too: $g(x(0), u(0), t) = 0$. The case of SG-algorithms in finite form and the case of inequality constraints are considered in a similar way.

The speed-gradient algorithm is tightly associated to the concept of Lyapunov function $V(x)$ – a function of the system state nonincreasing or nondecreasing along its trajectories. Lyapunov function is an abstraction for such physical characteristics as energy and entropy. In nonlinear dynamical systems theory it provides a powerful tool for analyzing stability-like properties [227, 316, 380, 419]. It is important that Lyapunov function can be used not only for analysis but also for system design. In particular, for the speed-gradient algorithms in the finite form the goal function itself may serve as the Lyapunov function : $V(x) = Q(x)$. The Lyapunov function for differential form of SG-algorithms is as follows: $V(x, u) = Q(x) + 0.5(u - u_*)^{\mathrm{T}} \Gamma^{-1}(u - u_*)$, where u_* is the desired "ideal" value of controlling variables. Note that in order to justify discrete-time gradient algorithm one may use Lyapunov function as square distance between the current and the "ideal" controlling variables $V(u) = |u - u_*|^2$.

2.4.3 Feedback linearization

The gradient and the speed-gradient methods represent a family of *goal-oriented methods* which allow the designer of control system to create the control algorithm for a nonlinear controlled system as soon as the control goal is formulated by means of a goal function. However, it is not a unique approach to control system design. A number of other more sophisticated approaches can be found in control literature, see [157, 164, 210, 227, 245, 464]. Below, one of the most popular methods: *feedback linearization* will be briefly presented.

Consider the systems affine in control:

$$\dot{x} = f(x) + g(x)u. \qquad (2.4.55)$$

Definition 2.1. System (2.4.55) is called *feedback linearizable in the open domain* $\Omega \in \mathbb{R}^n$ if there exist a smooth coordinate change $z = \Phi(x)$, $x \in \Omega$ and a feedback transformation

$$u = \alpha(x) + \beta(x)v \qquad (2.4.56)$$

with smooth functions α, β such that Φ and β are smoothly invertible in Ω and the closed loop system (2.4.55)–(2.4.56) is linear, i.e., there exist constant matrices $A \in \mathbb{R}^{n \times n}$ and $B \in \mathbb{R}^{n \times m}$ so that

$$f(x) + g(x)\alpha(x) = A, \quad g(x)\beta(x) = B, \quad x \in \Omega. \tag{2.4.57}$$

Feedback linearizability of the system means that it is equivalent to the system

$$\dot{z} = Az + Bv, \tag{2.4.58}$$

where $z(t) \in \mathbb{R}^n$ is the new state vector and $v(t) \in \mathbb{R}^m$ is the new input, which contains the nonlinearities.

Definition 2.2. System (2.4.55) is said to have *relative degree* r, $r \leq n$ at point $x_0 \in \mathbb{R}^n$ with respect to the output

$$y = h(x), \tag{2.4.59}$$

if for any $x \in \Omega$, where Ω is some neighborhood of x_0, the following conditions are valid:

$$L_g L_f^k h(x) = 0, \quad k = 0, 1, \ldots, r-2, \quad L_g L_f^{r-1} h(x) \neq 0.$$

Recall that $L_\psi \phi(x) = \sum_{i=1}^n \frac{\partial \phi}{\partial x_i} \psi_i(x)$ stands for the Lie derivative of the vector function ϕ along the vector field ψ. Relative degree r is exactly equal to the number of times one has to differentiate the output in order to have the input explicitly appearing in the equation which describes the evolution of $y^{(r)}(t)$ in the neighborhood of x_0.

Theorem 2.1 (Criterion of feedback linearizability for single-input /single-output systems). *System (2.4.55) is feedback linearizable in the neighborhood Ω of a point $x_0 \in \mathbb{R}^n$ if and only if there exists a smooth scalar function $h(x)$ defined in Ω such that the relative degree r of (2.4.55), (2.4.59) is equal to n.*

In the case $r = n$ the state transformation $z = \Phi(x)$ and the feedback law reducing (2.4.55) to the chain of integrators $\dot{y}_1 = y_2, \dot{y}_2 = y_3, \ldots, \dot{y}_n = u$ (so called *Brunovsky form*) can be chosen as follows:

$$\Phi(x) = \text{col}(h(x), L_f h(x), \ldots, L_f^{n-1} h(x)) \tag{2.4.60}$$

$$u = \frac{1}{L_g L_f^{n-1} h(x)} \left[-L_f^n h(x) + v \right]. \tag{2.4.61}$$

Example 2.1. Consider a simple pendulum without friction

$$\ddot{\varphi} + \sin \varphi = u. \tag{2.4.62}$$

The system (2.4.62) can be transformed to the form (2.4.55) by introducing the state vector $x = \text{col}(\varphi, \dot{\varphi}) \in \mathbb{R}^2$, i.e., $x_1 = \varphi$, $x_2 = \dot{\varphi}$. In this case $f(x) = \text{col}(x_2, -\sin x_1)$, $g(x) = \text{col}(0, 1)$. Choose $y = \varphi$, i.e., $h(x) = x_1$. Then

$$L_g h(x) = [0\ 1][0\ 1]^T = 0, \quad L_f h(x) = [1\ 0]f(x) = x_2, \quad L_g L_f h(x) = 1.$$

Therefore, $r = 2$ which means that the system (2.4.62) is feedback linearizable. The linearizing feedback (2.4.61) is as follows

$$u = -\sin x_1 + v. \qquad (2.4.63)$$

This feedback reduces the system (2.4.62) to the form of double integrator

$$\dot{x}_1 = x_2, \quad \dot{x}_2 = v. \qquad (2.4.64)$$

Note that although the system is linearized by feedback, it is not stabilized yet! If our primary goal is stabilization of the system at the point x_0 we still need more effort to achieve it. It is more or less clear that the system (2.4.64) cannot be asymptotically stabilized at the origin by a smooth output feedback. On the other hand the stabilization problem can be easily solved if the second state variable x_2 is available for measurement. In this case the stabilizing feedback is $v = -\mu x_2$ for any $\mu > 0$, or $u = -\mu x_2 - \sin x_1$. However, for higher order systems, introducing the derivatives into the feedback law may be hard to implement.

2.5 Results: Laws of cybernetical physics

According to the previous discussion, a typical control problem is to find a control function (feedback operator) from the given class of functions (operators) ensuring the given control goal for the given class of systems. However, for physics or other natural science such a formulation may play only an auxillary role, since a control goal is rarely specified *a priori*. For physics it would be more natural to find what properties or what behaviors of a system can be achieved or changed by means of applying control function from the given class. How to express solutions to such problems in a manner that physicists are more accustomed to?

A great deal of the results in many areas of physics are presented in the form of *conservation laws*, stating that some quantities do not change during evolution of the system. However, the formulations in cybernetical physics are different. Since the results in cybernetical physics establish how the evolution of the system can be changed by control, they should be formulated as *transformation laws*, specifying the classes of changes in the evolution of the system attainable by control function from the given class, i.e., specifying the limits of control.

Let us provide a few examples of transformation laws. The first example is related to control of an invariant (constant of motions) for a conservative system. In this case transformation law should provide an answer to the question: "What can be done with an invariant by means of a feedback?" A typical result (see, e.g., [407] and Chapter 3) can be loosely formulated as follows:

> The value of any controllable invariant can be changed for arbitrary quantity by means of an arbitrarily small feedback.

2.5 Results: Laws of cybernetical physics

The meaning of the term "controllable" depends on a specific situation and, in principal, includes some conditions ensuring solvability of the problem. Examples will be given in Chapter 3.

The second transformation law relates to dissipative systems. The results, presented in Chapter 4 demonstrate that the smaller the dissipation in the system the larger efficiency of a small feedback. A typical quantitative result may be expressed as follows:

> The level of energy achievable by means of control of the power γ for controllable Hamiltonian or Lagrangian system with small dissipation of the degree ρ has the order $(\gamma/\rho)^2$.

The third example of transformation law relates to control of chaos. It was first articulated in the seminal paper [331] and can be termed the *OGY-law*:

> Any controllable chaotic trajectory can be transformed into a periodic one by means of an arbitrarily small control.

Note that the chaoticity requirement can be significantly weakened, being replaced by some form of *recurrency*. Again the term "controllable" in the above context means principal solvability of the problem.

To provide more elaborated formulation of the law some sufficient conditions ensuring controllability may be checked or imposed. It may be a matter of further mathematical investigation; some results will be presented in the following chapters of this book.

Summarizing this chapter, note that the subject of cybernetical physics includes system models and control goals. Its problems are studying behavior of physical systems under external (feedforward or feedback) purposeful actions. So, defined subject has many similarities with that of the theory of open systems [236]. The main difference is that feedbacks are not prespecified and need to be designed.

The methodology of feedback design is borrowed from cybernetics (control theory). It is based upon methods of controlled system model's building, methods of system states and parameters estimation (identification), and methods of feedback synthesis. The models of controlled system used in cybernetics differ from conventional models of physics and mechanics in that they have explicitly specified inputs and outputs. Unlike conventional physics results, often formulated as conservation laws, the results of cybernetical physics are formulated in the form of transformation laws, establishing the possibilities and limits of changing properties of a physical system by means of control.

Thus, in this new research area a synthesis of descriptive and prescriptive sciences is accomplished. Such an extension of conventional physical study, scope, and methodology helps to achieve its ultimate goal: better understanding of the Nature.

3
Control of Conservative Systems

In a number of control problems in physics the control goal is expressed in terms of a quantity that is invariant (first integral) of an uncontrolled system. The most important class of such problems is control of energy where the control goal can be formulated in terms of the system energy. Below formal statements of energy control problems and designs of control algorithms are presented for systems with Hamiltonian and Lagrangian dynamics. Control algorithms are proposed using the speed-gradient method. Mathematical conditions ensuring achievement of the control goal with arbitrarily small intensity of control and extensions to control of several first integrals are given.

3.1 Control of energy for Hamiltonian systems

Hamiltonian formalism is used in physics to describe dynamics of various systems, from motion of planets to motion of molecules. It is also a convenient mathematical description for controlled oscillatory systems, since it allows for explicit description of surfaces of constant energy which unforced oscillatory motions belong to. Energy is the unforced system invariant (first integral, constant of motion), and a measure of interaction of the system with its environment.

The problem of controlling the change of the system energy is of utmost importance. Not only it has a fundamental theoretical value, it also allows a number of practical problems related to energy saving technologies to be solved. Therefore, we start studying control problems in physics from control of energy based on Hamiltonian description of system dynamics.

Assume that the system is conservative: dissipation and losses are negligible. The Hamiltonian form of controlled system equations is as follows:

$$\dot{p}_i = -\frac{\partial H(p, q, u)}{\partial q_i}, \qquad \dot{q}_i = \frac{\partial H(p, q, u)}{\partial p_i}, \quad i = 1, \ldots, n, \qquad (3.1.1)$$

where $p = \mathrm{col}(p_1, \ldots, p_n)$, $q = \mathrm{col}(q_1, \ldots, q_n)$ are the vectors of generalized coordinates and momenta constituting the state vector $\mathrm{col}(p,q)$ of the system. $H = H(p,q,u)$ is the controlled Hamiltonian function, $u(t) \in \mathbb{R}^m$ is the input (generalized force).[1] The model (3.1.1) can be rewritten as follows

$$\begin{cases} \dot{p} = -\nabla_q H(p, q, u), \\ \dot{q} = \nabla_p H(p, q, u). \end{cases} \quad (3.1.2)$$

Following [138] the control goal is formalized as approaching a given energy surface of the free (unforced) system

$$\mathrm{col}(p(t), q(t)) \to S, \quad (3.1.3)$$

where $S = \{(p,q) : H_0(p,q) = H_*\}$, $H_0(p,q) = H(p,q,0)$ is the "internal" Hamiltonian describing the unforced system

$$\begin{cases} \dot{p} = -\nabla_q H_0(p, q), \\ \dot{q} = \nabla_p H_0(p, q). \end{cases} \quad (3.1.4)$$

A slightly different formulation of control goal is the following

$$H_0(p(t), q(t)) \to H_*, \quad \text{when} \quad t \to \infty. \quad (3.1.5)$$

The goal (3.1.5) means convergence of the value of the system energy to the desired value H_*. Formally, is not equivalent to (3.1.3), meaning convergence of the arguments of the energy function. Indeed, (3.1.3) implies (3.1.5) but (3.1.5) does not imply (3.1.3) if S is not compact. Conversely, (3.1.5) implies (3.1.3) for Lagrangian systems with uniformly positive definite inertia matrix. In this chapter the mathematical difference between (3.1.5) and (3.1.3) will be ignored which influences neither physical nor engineering applications.

Introduce the following objective function

$$Q(x) = \frac{1}{2}(H_0(p,q) - H_*)^2, \quad (3.1.6)$$

where $x = \mathrm{col}(p, q)$. Then the control goal (3.1.5) takes the form

$$Q(x(t)) \to 0 \quad \text{when} \quad t \to \infty. \quad (3.1.7)$$

In what follows we assume that the Hamiltonian is linear in control:

$$H(p, q, u) = H_0(p, q) + H_1(p, q)^\mathrm{T} u,$$

where $H_0(p,q)$ is the internal Hamiltonian and $H_1(p,q)$ is an m-dimensional vector (column) of interaction potentials [316].

---

[1] Hereafter the variables are assumed to be dimensionless quantities if the opposite is not specified.

3.1 Control of energy for Hamiltonian systems

Define the Poisson bracket of smooth functions $f(p,q)$ and $g(p,q)$ in a standard manner

$$[f,g] = \sum_{i=1}^{n} \left(\frac{\partial f}{\partial p_i} \frac{\partial g}{\partial q_i} - \frac{\partial f}{\partial q_i} \frac{\partial g}{\partial p_i} \right).$$

If f and g are the vector-functions then the Poisson bracket is defined componentwise. For example, if the function f is scalar and g is an m-dimensional vector (column) then $[f,g]$ is an m-dimensional co-vector (row). More generally, if f and g are an l-dimensional and m-dimensional vectors, respectively, then $[f,g]$ is an $l \times m$ matrix.

Now let us apply the speed-gradient (SG) method described in Chapter 2 to solve the posed problem. To design the SG algorithm calculate $\dot{Q}$:

$$\dot{Q} = (H_0 - H_*) \left(\frac{\partial H_0}{\partial p} \dot{p} + \frac{\partial H_0}{\partial q} \dot{q} \right) = (H_0 - H_*)[H_0, H_1^{\mathrm{T}}]u \qquad (3.1.8)$$

and the speed-gradient: $\nabla_u \dot{Q} = (H - H_*)[H_0, H_1]$.

The differential SG algorithm can be represented in the form

$$\dot{u} = -\gamma (H_0 - H_*)[H_0, H_1], \qquad (3.1.9)$$

while the linear and relay finite forms (2.4.48), (2.4.51) are as follows:

$$u = -\gamma (H_0 - H_*)[H_0, H_1], \qquad (3.1.10)$$

$$u = -\gamma \mathrm{sign}\left\{ (H_0 - H_*)[H_0, H_1] \right\}, \qquad (3.1.11)$$

where $\gamma > 0$ is the gain factor. We may consider also the general speed-pseudogradient algorithm

$$u = -\psi \left((H_0 - H_1)[H_0, H_1] \right), \qquad (3.1.12)$$

where ψ is a smooth vector function with values in $\mathbb{R}^m$ which satisfies the strict pseudogradient condition $\psi(z)^{\mathrm{T}} z > 0$ for $z \neq 0$.

To analyze the behavior of the system with algorithms (3.1.9)–(3.1.11) we need conditions guaranteeing achievement of the control goal. Let us establish such conditions.

Theorem 3.1. [164]. *Let the first and second derivatives of the functions H_0, H_1 be bounded on the set $\Omega_0 = \{x : Q(x) \le Q_0\}$ for some $Q_0 > 0$.*

Then the algorithm (3.1.12) with $x(0) \in \Omega_0$ ensures $u(t) \to 0$ when $t \to \infty$ and ensures either the goal (3.1.5) or convergence $[H_0, H_1](x(t)) \to 0$ when $t \to \infty$.

Let, additionally, the following two conditions hold:

H1. *For any $c \ne H_*$ there exists $\varepsilon > 0$ such that any nonempty connected component of the set $D_{\varepsilon,c} = \{x : |[H_0(x), H_1(x)]| \le \varepsilon, \ |H_0(x) - c| \le \varepsilon\} \cap \Omega_0$ is bounded.*

H2. *The largest invariant set $M \subset D_0$ of the free system (i.e., the set M of whole trajectories of (3.1.4) contained in D_0), where $D_0 = \{x : [H_0(x), H_1(x)] = 0\} \cap \Omega_0$, consists of finite or countable number of isolated points.*

Then any solution of the system (3.1.2), (3.1.12) either achieves the goal (3.1.5) or tends to a point of the set D_0 which is an equilibrium of the free system (3.1.4). Besides, the set of initial conditions from which the solution of (3.1.2), (3.1.12) tends to unstable[2] equilibrium of the free system has the zero Lebesgue measure.

Corollary 3.1. *If D_0 is empty, i.e., $[H_0, H_1](x) \neq 0$ for $x \in \Omega_0$, then the control goal is achieved for all $x(0) \in \Omega_0$.*

Proof of Theorem 3.1. Calculation of the time derivative of $Q_t = (H_0(p(t), q(t)) - H_*)^2/2$ along the solution of (3.1.2), (3.1.12) yields

$$\dot{Q}_t = -\psi(z(t))^\mathrm{T} z(t), \qquad (3.1.13)$$

where $z(t) = [H_0(x(t)), H_1(x(t))]^\mathrm{T}(H_0(x(t)) - H_*)$. Hence $\dot{Q}_t \leq 0$ and Q_t does not increase, i.e., $Q_t \leq Q_0$. It means that the solution of the whole system will never leave the set Ω_0. Boundedness of the right-hand side of the system (3.1.2), (3.1.12) ensures that $x(t)$ is well defined for all $0 \leq t < \infty$. Therefore, there exist $\lim_{t\to\infty} Q_t = Q_\infty$ and $\lim_{t\to\infty} H_0(x(t)) = H_\infty$. If $H_\infty = H_*$ then the theorem is proved.

Suppose $H_\infty \neq H_*$. The boundedness condition gives that $z(t)$, $\psi(z(t))$, $\dot{Q}_t$ and $\ddot{Q}_t$ are bounded for $x(t) \in \Omega_0$. It follows from the Partial Stability Theorem [157] that $\dot{Q}_t \to 0$. By virtue of the strict pseudogradient condition and continuity of ψ one can deduce that $u(t) \to 0$ and $z(t) \to 0$ when $t \to \infty$. By assumption $H_\infty \neq H_*$. Hence $[H_0, H_1](x(t)) \to 0$. The first part of the theorem is proved.

To prove the second part choose $\varepsilon > 0$ from Assumption H2. For sufficiently large $t > 0$ the solution $x(t)$ enters the set $D_{\varepsilon, H_\infty}$ and does not leave one of its connected components which is bounded and therefore its closure is compact. By virtue of compactness there exists a limit point of $x(t)$ and all limit points of $x(t)$ satisfy $[H_0, H_1] = 0$, i.e., $x(t)$ converges to the set $D_0 = \{x : [H_0, H_1] = 0\} \cap \Omega_0$. According to the LaSalle theorem (see, e.g., [157, 380]) $x(t)$ tends to the largest invariant set of the free system in D_0. Taking into account condition H2, we conclude that there exists $\lim_{t\to\infty} x(t) = x_\infty \in D_0$ and x_∞ is an equilibrium point of (3.1.2), (3.1.12). Let A_∞ be the Jacobi matrix calculated at the point x_∞ and M_s, M_u, M_0 be stable, unstable, and central manifolds of the system in x_∞. Then it follows from the Center Manifold Theorem [189, 210] that $x(t) \to x_\infty$ only for $x(0) \in M_s \oplus M_0$ if $|x(0) - x_\infty|$ is sufficiently small. Let x_∞ be an unstable equilibrium, i.e., M_u is not empty. Then $\dim M_s \oplus M_0 < 2n$. Making backward shift along trajectories of the free system, it is easy to show that all initial

---

[2] Instability of an equilibrium is understood here in the sense of mechanics, i.e., it means that the Jacobi matrix of the system calculated at the equilibrium point has at least one eigenvalue with positive real part.

conditions $x(0)$ such that $x(t) \to \infty$ belong to a some manifold of dimension less than $2n$. Since the set of all possible unstable limit points is either finite or countable, the set of corresponding initial conditions has the zero Lebesgue measure in $\mathbb{R}^{2n}$. ∎

The proved theorem shows that algorithm (3.1.12) ensures the goal (3.1.5) almost always unless there are "false" goals: stable or neutral equilibria of the free system which are reachable from the initial point within the energy layer

$$\Omega_0 = \{(p, q) \ : \ |H_0(p, q) - H_*| \leq |H_0(p(0), q(0)) - H_*|\}.$$

In other words, the goal (3.1.5) will be achieved for almost all initial conditions from the set Ω_0 if it does not contain local potential wells.

Moreover, it is clear from the proof that the set of exceptional initial conditions is contained within a finite or countable number of manifolds, i.e., the complement of this set is open dense in the set Ω_0.

Remark 3.1. Suppose that the Hamiltonian Description of the controlled system (3.1.2) originates from controlled Euler–Lagrange equations, i.e., its internal Hamiltonian and interaction potentials are

$$H_0(p, q) = \frac{1}{2} p^T A^{-1}(q) p + \Pi(q), \qquad H_1(p, q) = q, \qquad (3.1.14)$$

where $p(t), q(t) \in \mathbb{R}^n$, $A(q)$ is the positive definite matrix of kinetic energy and $\Pi(q)$ is the function of potential energy. In this case, equilibria of the unforced system have the form $(0, \widehat{q})$, where $\widehat{q}$ is a stationary (critical) point of the potential $\Pi(q)$. Suppose that all the stationary points of $\Pi(q)$ are isolated. Then it follows from Theorem 3.1 that almost all solutions of the closed loop system (3.1.2), (3.1.12) satisfy the goal (3.1.5). In addition, if the matrix $A(q)$ is uniformly positive definite, i.e., $z^T A(q) z \geq \mu |z|^2$ for some $\mu > 0$ and all $q \in \mathbb{R}^n$, $z \in \mathbb{R}^n$, then it is easy to show that almost all solutions of the system approach the goal set $S = \{(p, q) \ : \ H_0(p, q) = H_*\}$. □

Remark 3.2. The sign control algorithm

$$u = -\gamma \text{sign}\left\{(H_0 - H_*)[H_0, H_1]\right\} \qquad (3.1.15)$$

formally does not satisfy the conditions of the previous theorem because of the discontinuity of the function ψ in this case. However, the similar result can be proved for the system (3.1.2), (3.1.15), except for the property $u(t) \to 0$. Additionally, it can be proved that if the goal surface S is compact, then solutions of the system (3.1.2), (3.1.15) achieve it in a finite time. □

3.2 Example: Controlled pendulum

In order to demonstrate application of the obtained result, consider a simple pendulum model (see Chapter 1). It can be transformed into the Hamiltonian form with the energy (Hamiltonian) function

$$H_0(p,q) = \frac{p^2}{2J} + mgl(1 - \cos q), \qquad (3.2.16)$$

where $q(t) \in \mathbb{R}^1$ is the angular coordinate, $p = J\dot{q}$ is the momentum of the system. The Hamiltonian form of the controlled system can be written

$$\begin{cases} \dot{q} = J^{-1}p \\ \dot{p} = -mgl\sin q + u(t), \end{cases} \qquad (3.2.17)$$

where $u(t)$ is the controlling torque. In this example the interaction potential has the form $H_1(p,q) = q$. The goal (3.1.5) corresponds to swinging the pendulum up or down to the amplitude

$$q_* = \arccos\left(1 - \frac{H_*}{mgl}\right) \quad \text{for} \quad 0 \le H_* < 2mgl.$$

In the case of larger desired energy level: $H_* > 2mgl$ the goal corresponds to the rotatory motion of the pendulum. The value $H_* = 2mgl$ is exceptional. It corresponds to a motion along separatrix – the set consisting of a number of smooth curves separating domains of oscillatory and rotatory motions in the phase plane. Speed-gradient algorithms (3.1.10), (3.1.11) for pendulum (3.2.17) take a very simple form:

$$\dot{u} = -\gamma(H_0 - H_*)\dot{q}, \qquad (3.2.18)$$

$$u = -\gamma \operatorname{sign}\left((H_0 - H_*)\dot{q}\right). \qquad (3.2.19)$$

Applying Theorem 3.1 and Remark 3.1 to the pendulum (3.2.17) it can be concluded that if the initial energy layer between the levels $H_0(p(0), q(0))$ and H_* does not contain an equilibrium of the unforced system, then the goal level H_* will be achieved in the controlled system (3.1.2), (3.1.12) from all initial conditions. If the initial energy layer contains only unstable equilibria $(\pi(2k+1), 0)$, $k = \pm 1, \pm 2, \ldots$, then the goal (3.1.5) will be achieved from almost all initial conditions.

3.3 The swinging (small control) property

Most applications of oscillation control systems require the control action to be small. The reason is that the lifetime of the system is usually large compared with the typical period of oscillations, and therefore the control power (or energy spent over one cycle) should be small.

This gives rise to the problem: What kind of control goals can be achieved by small control? For problems of control of Hamiltonian systems the question is: What values of energy H_* can be achieved by small control? Keeping in mind the problem of swinging a simple pendulum the following terminology is introduced.

Definition 3.1. The system
$$\dot{x} = F(x, u, t)$$
is called *swingable with respect to the goals*
$$\lim_{t \to \infty} Q_t = g_*, \quad g \in G \subset \mathbb{R}^1, \tag{3.3.20}$$
if for any $\varepsilon > 0$ and any $g \in G$ there exists the control law
$$u(t) = \mathcal{U}_{g,\varepsilon}[x(s), \ 0 \le s \le t], \tag{3.3.21}$$
such that $|u(t)| < \varepsilon$ and for the closed loop system the goal (3.3.20) is achieved. The control law (3.3.21) in that case is called swinging control with respect to G.

It follows from Theorem 3.1 that any Hamiltonian controlled system satisfying H1, H2 is swingable for almost all initial states (p,q) if its potential $\Pi(q)$ has only isolated stationary points and $g \ge \hat{g}$, where $\hat{g} = \sup_i[\Pi(q^i)]$ and q^i are local minima of $\Pi(q)$. Under the conditions of Theorem 3.1, the swinging control can be determined by Eq. (3.1.10) for
$$\gamma < \varepsilon(h|H_0(p(0), q(0)) - H_*|)^{-1},$$
where
$$h = \sup_{\Omega_0} |[H_0(p,q), H_1(p,q)]|,$$
$$\Omega_0 = \{(p,q) : |H_0(p,q) - H_*| \le |H(p(0), q(0)) - H_*|\}.$$
If there are no equilibria of the free system in the set Ω_0, then the goal is achieved for any trajectory of the system.

The above result can be applied to a variety of control problems for oscillatory systems. Some examples will be given in the subsequent chapters. Note that the smaller the level of control the longer the *transient time* required to achieve the goal. Transient time is important for engineering and other practical applications. However, the possibility to achieve the goal in principle is important for evaluation of limits of control.

3.4 Control of first integrals

One may wonder if it is possible to achieve more complex goals than approaching a given energy surface. The natural extension of the control goal (3.1.5) could be achieving the desired level of several first integrals (conserved quantities) of the unforced system. The speed-gradient algorithms apply to this more general situation as well. In this section we formulate a formal result following [158].

Let $x = \mathrm{col}(p, q) \in \mathbb{R}^{2n}$ be the state vector of the Hamiltonian system (3.1.2) and the smooth functions $G_i : \mathbb{R}^{2n} \to \mathbb{R}^1$, $i = 1, \ldots, k$ be the first integrals of the free system (3.1.2), i.e.,

$$[H_0, G_i] \equiv 0, \quad i = 1, \ldots, k, \tag{3.4.22}$$

where $H_0(p, q) = H(p, q, 0)$. Let the control goal be

$$G_i(x(t)) \to G_i^*, \quad i = 1, \ldots, k, \quad \text{when } t \to \infty \tag{3.4.23}$$

where $G_i^*, i = 1, \ldots, k$ are prespecified numbers.

Assume again that the Hamiltonian is affine in control (although the results can be extended to the nonaffine case, see [404]) and consider the system

$$\begin{aligned} \dot{q} &= \nabla_p H_0(q, p) + \sum_{j=1}^m \nabla_p H_j u_j, \\ \dot{p} &= -\nabla_q H_0(q, p) - \sum_{j=1}^m \nabla_q H_j u_j, \end{aligned} \tag{3.4.24}$$

where H_0 is the Hamiltonian function of the unforced system (3.4.24); $H_j, j = 1, \ldots, m$ are the interaction potentials being independent functions (in the sense that the corresponding one-forms dH_j are linearly independent) [316]; $u_j, j = 1, \ldots, m$ are controlling inputs of the system.

To solve the posed problem we again use the speed-gradient method and construct the following control algorithm

$$u(q(t), p(t)) = -\gamma \nabla_u \dot{Q}(q(t), p(t)), \tag{3.4.25}$$

or, more generally

$$u(q(t), p(t)) = -\psi(\nabla_u \dot{Q}(q(t), p(t))), \tag{3.4.26}$$

where $Q(q, p)$ is the goal functional, $\dot{Q}$ is the full derivative of $Q(q, p)$ along the solutions of (3.4.24) and $\psi(z)$ is a vector-function forming an acute angle with z, i.e., $\psi(z)^\mathrm{T} z > 0$ for $z \neq 0$. Take the goal functional as follows

$$Q(q, p) = \frac{1}{2}(G(q, p) - G^*)^\mathrm{T} R (G(q, p) - G^*), \tag{3.4.27}$$

where $G(q, p) = (G_1(q, p), \ldots, G_k(q, p))^\mathrm{T}$, $G^* = (G_1^*, \ldots, G_k^*)^\mathrm{T}$ and R is symmetric positive definite constant matrix. Then the corresponding speed-gradient control algorithm (3.4.25) has the following form

$$u = -\gamma[\bar{H}, Q] = -\gamma[\bar{H}, G](q, p) R(G(q, p) - G^*), \tag{3.4.28}$$

where $\bar{H}$ stands for the column vector with components H_j.

The general algorithm (3.4.26) is as follows

$$u = -\psi([\bar{H}, Q]). \tag{3.4.29}$$

Introduce the set

$$S(q,p) = \text{span}\{\text{ad}_{H_0}^s[\bar{H},G],\ s=0,1,\ldots\}.$$

where for every $H, G \in \mathcal{C}^\infty$ we define inductively $\text{ad}_H^0 G = G$, $\text{ad}_H^1 G = [H,G]$, $\text{ad}_H^{s+1} G = [H, \text{ad}_H^s G]$.

Theorem 3.2. [158, 164]. *Consider the controlled Hamiltonian system (3.4.24) defined on a smooth $2n$-dimensional manifold M^{2n} with Hamiltonians H_0, H_j bounded together with their first and second partial derivatives in the set $\Omega_0 = \{(p,q) : Q(p,q) < \varepsilon\}$ for some $\varepsilon > 0$. Let $G_i, i=1,\ldots,k$ be $\mathcal{C}^\infty$-smooth conserved quantities of the unforced system (3.4.24). Assume that there exists $\delta > 0$ such that each connected component of the set*

$$D_\delta = \Omega_0 \cap \{(p,q) : \det A^\mathrm{T} A \leq \delta\}$$

is bounded, where $A = [\bar{H}, F]$ and

$$\dim S(q,p) \geq k \quad \forall (q,p) \in \Omega_0, \tag{3.4.30}$$

Then the control goal (3.4.23) is achieved for any trajectory of the system (3.4.24), (3.4.29) with the initial conditions from the set Ω_0.

Remark 3.3. The condition (3.4.30) was introduced by A. Shiriaev [158, 164]. It is of utmost importance for the achievement of the goal set. In case when the goal set is a single point Shiriaev's condition (3.4.30) implies zero-state detectability[3] which, in turn ensures stabilization of the origin [96]. In general case the condition (3.4.30) can be thought of as a set detectability condition or condition of controllability with respect to the vector output G. □

Remark 3.4. If the condition (3.4.30) holds everywhere in Ω_0 except some set M_0 of isolated points (p_*, q_*) then $(p_*, q_*) \in M_0$ is an equilibrium of the unforced system. It follows from the Center Manifold Theorem [189, 210] that, if the unforced system has only isolated equilibria (p_*, q_*) in Ω_0 and each of them is hyperbolically unstable (in sense that the corresponding Jacobi matrix has at least one eigenvalue with the positive real part) then the Lebesgue measure of the initial conditions for which the control goal is not achieved is equal to zero. □

In fact, Theorem 3.2 states that, if we have avoided convergence to the stable equilibrium, then the control goal will be achieved for almost all initial conditions under the controllability-like condition (3.4.30).

When the goal functions Q are radially unbounded, the explicit conditions ensuring the goal (3.4.23) can be given.

Corollary 3.2. *Let Q be a radially unbounded function, i.e., the set $\{(p,q) : Q(p,q) \leq c\}$ is compact for all $c \in \mathbb{R}^1$. Let $\dim S(p,q) \geq k$ for all $(p,q) \in \Omega_0$,*

---

[3] The system $\dot{x} = F(x), y = h(x)$ is called *zero-state detectable* if the relation $y(t) \to 0$ when $t \to \infty$ implies $x(t) \to 0$ when $t \to \infty$.

where $\Omega_0 = \{(p,q) : Q(p,q) \leq Q_0\}$. Then the goal (3.4.23) is achieved in the system (3.4.24), (3.4.28) for all initial conditions $(p(0), q(0)) \in \Omega_0$.

Remark 3.5. The simple condition eliminating convergence to a stable equilibrium is just the absence of stable equilibria in the connected component of the set $\{(q,p) : Q(q,p) \leq Q(q(0), p(0))\}$. To satisfy it the proper choice of the goal function $Q(q,p)$, i.e., the proper choice of the values G_i^* and the weighting matrix R, may help. □

Remark 3.6. In case when $k = n$ and each pair of the functions G_i are in involution, i.e.,
$$[G_i, G_j] = 0, \quad i,j = 1, \ldots, k, \qquad (3.4.31)$$
it can be proved that each solution of the closed loop system tends to some solution of the unforced system which is either a quasiperiodic one, or an equilibrium point [37]. Hence, the behavior of the closed loop control system (3.4.24), (3.4.29) cannot be chaotic. □

3.5 Control of generalized Hamiltonian systems

The proposed approach applies also to the so-called generalized Hamiltonian systems [390] which can be described in the canonical local coordinates as follows
$$\begin{cases} \dot{q} = \nabla_p H_0(q,p,s) + g_q(q,p,s)u, \\ \dot{p} = -\nabla_q H_0(q,p,s) + g_p(q,p,s)u, \\ \dot{s} = g_s(q,p,s)u, \end{cases} \qquad (3.5.32)$$

where $q(t) \in \mathbb{R}^n, p(t) \in \mathbb{R}^n, s(t) \in \mathbb{R}^l$ are the state variables and H_0, g_q, g_p, g_s are some smooth functions.

Obviously, the function H_0 in (3.5.32) is an invariant of the unforced system
$$\begin{cases} \dot{q} = \nabla_p H_0(q,p,s), \\ \dot{p} = -\nabla_q H_0(q,p,s), \\ s = \text{const.} \end{cases} \qquad (3.5.33)$$

Suppose that some set of the invariants $H_1, \ldots, H_m$ of the unforced system is given. Then we may pose the problem of achieving the goal
$$\lim_{t \to \infty} H_i(q(t), p(t), s(t)) = H_{i*}, \quad i = 1, \ldots, k \qquad (3.5.34)$$

and design the speed-gradient algorithm (3.4.26) as follows. Choose the partial goal functional
$$Q_i = (H_i - H_{i*})^2/2.$$
Then calculate its derivative along (3.5.32)

3.5 Control of generalized Hamiltonian systems

$$\dot{Q}_i = (H_i - H_{i*}) \left[\frac{\partial H_i}{\partial p} \dot{p} + \frac{\partial H_i}{\partial q} \dot{q} + \frac{\partial H_i}{\partial s} \dot{s} \right]$$

and the speed-gradient

$$\nabla_u \dot{Q}_i = (H_i - H_{i*}) \left[\frac{\partial H_i}{\partial p} g_p + \frac{\partial H_i}{\partial q} g_q + \frac{\partial H_i}{\partial s} g_s \right]^{\mathrm{T}}.$$

Therefore the algorithm (3.4.25) reads

$$u = -\gamma \sum_{i=1}^{m} \nabla H_i^{\mathrm{T}} g (H_i - H_{i*}), \qquad (3.5.35)$$

where $g = \mathrm{col}(g_q, g_p, g_s)$. The conditions which guarantee achievement of the goal (3.5.34) look similar to those of Theorem 3.2.

A special case of (3.5.32) is a mechanical system with kinematic constraints. Consider the Lagrange–Euler system with the Hamiltonian

$$H = \frac{1}{2} \dot{q}^{\mathrm{T}} M(q) \dot{q} + \Pi(q), \qquad (3.5.36)$$

where $M(q)$ is the positive definite matrix of kinetic energy, $\Pi(q)$ is potential energy. Suppose there are k kinematic constraints on the generalized velocities: $A(q)^{\mathrm{T}} \dot{q} = 0$, where the $k \times n$ matrix $A(q)$ has rank k. Applying the elimination procedure for k-dependent generalized coordinates (see [390]) we arrive at a generalized Hamiltonian description in the space of reduced dimension $2n - k$. The control algorithm ensuring the goal (3.5.34) can be derived similarly to (3.5.35).

The results presented in this section establish the possibilities as well as some of the limitations of SG algorithms for organizing oscillatory behavior of nonlinear Hamiltonian systems. The proposed algorithms ensure the control goal for arbitrary G_i^*, and therefore for arbitrary energy levels of the system. Moreover, the goal can be achieved with arbitrary small $\gamma > 0$, i.e., for an arbitrary low control level (swinging property). The results have been extended to the generalized Hamiltonian systems and systems with constraints.

It is interesting to compare the above results with the KAM-theory [37] which in essence analyzes the behavior of a system with the uncontrolled perturbed Hamiltonian. One of the core results of the KAM-theory can be interpreted as follows: the perturbed system with the Hamiltonian $H_\varepsilon(q, p) = H_0(q, p) + \varepsilon H_d(q, p)$ generically becomes chaotic as ε grows. Our results show that the controlled perturbed system with the Hamiltonian $H(q, p) = H_0(q, p) + \sum_{j=1}^{m} H_j(q, p) u_j$ and SG feedback with $k = m = n$ will never create chaos since the trajectories of the closed loop system for arbitrary gain tend to quasiperiodic motions.

Note that if $m < n$, i.e., the system is "underactuated," then the SG algorithm may produce chaos even for the case of energy control problem for double pendulum ($n = 2, m = 1$) [240]. Moreover, as shown in [240] by numerical computer experiments the change of the goal energy value H_* leads to the change of the upper Lyapunov exponent of the closed system in a wide range.

4

Control of Dissipative Systems

For control of energy in the case of system with dissipation the limits of energy transformation by means of control are established. To this end the notion of excitability index is introduced and analyzed. The resonance-like behavior of systems under feedback excitation (phenomenon of feedback resonance) is studied.

4.1 Excitability analysis of dissipative systems

Let us study energy control problems for systems with dissipation. Departing from the Hamiltonian description consider a system with dissipation modeled as follows:

$$\dot{q}_i = \frac{\partial H(q,p,u)}{\partial p_i}, \quad \dot{p}_i = -\frac{\partial H(q,p,u)}{\partial q_i} - R_i(q,p), \quad i = 1,\ldots,n, \quad (4.1.1)$$

where $q = \mathrm{col}(q_1,\ldots,q_n)$, $p = \mathrm{col}(p_1,\ldots,p_n)$ are vectors of generalized coordinates and generalized momenta forming the state vector of the system $x = \mathrm{col}(q,p)$; $H = H(q,p,u)$ is the Hamiltonian of the controlled system; $u(t) \in \mathbb{R}^m$ is input (vector of external generalized forces); $R(q,p) = \mathrm{col}(R_1(q,p),\ldots,R_n(q,p))$ is the dissipation function satisfying the inequality

$$R(q,p)^\mathrm{T} \frac{\partial H_0(q,p)}{\partial p} \geq 0, \quad (4.1.2)$$

where $H_0(q,p) = H(q,p,0)$ is the energy of the free system. The fulfillment of inequality (4.1.2) means dissipation of energy along the trajectories of the free system: $\dot{H}_0 \leq 0$.

In the case when the Hamiltonian is affine in control: $H(q,p,u) = H_0(q,p) + H_1(q,p)u$ the dissipation inequality is more explicit:

$$\dot{H}_0 = [H_0, H_1]u - R(q,p)^\mathrm{T} \frac{\partial H_0(q,p)}{\partial p} \quad (4.1.3)$$

4 Control of Dissipative Systems

Apparently, dissipation complicates the control of the system energy, especially energy pumping and makes swingability of the system infeasible. It is of interest to evaluate the limits of possible energy change for given levels of control and dissipation. Such a limit would provide a law of energy transformation under control action. The case of a small dissipation (weakly damped systems) is of special interest since in this case the system may exhibit oscillation and resonance phenomena. In order to evaluate limits of energy control, we introduce the measure of the system's ability to be excited by a bounded control: the *excitability index*. Then it will be shown how excitability index can be used for creation of resonance modes in nonlinear systems.

4.1.1 Excitability index

The measure of the system excitability depends on the choice of its input and output. In our case we study excitability with respect to the energy output of the system and examine the asymptotic (limit) value of the energy achievable in the system excited with a control of the given level γ. To realize maximum possible excitation, $u(t)$ should depend on the state of the system $x(t) = (q(t), p(t))$ or on the current measurements $y(t)$, which means introducing a state feedback $u(t) = U(x(t))$ or output feedback $u(t) = U(y(t))$. Now the problem is: how to find the feedback law in order to achieve the maximum limit amplitude of output? In [142] this problem was formulated as that of optimal control as follows. Find

$$\chi(\gamma) = \limsup_{\substack{|u(s)|\leq\gamma,\\ 0\leq s\leq t,\\ x(0)=0,\\ t\geq 0}} H_0(x), \qquad (4.1.4)$$

in the case when the limit in (4.1.4) exists, where $H_0(x) = H_0(q,p)$ is the energy of unforced system. In order to ensure $\chi(\gamma)$ to be well defined, it is assumed that the state of the system (4.1.7) is bounded for bounded inputs and $x = 0$ is an equilibrium of the unforced system ($\partial H_0/\partial p(0) = 0$, $\partial H_0/\partial q(0) = 0$, $H_0(0) = 0$). It is also assumed that energy function $H_0(x)$ is nonnegative.

It is well known in control theory that the signal providing a solution to an optimal control problem should depend not only on time but also on system state, i.e., input signal should have a feedback form. If the system (4.1.7) is linear, then due to superposition property the value of the problem (4.1.4) depends quadratically on γ. Since the energy is proportional to the squared amplitude of the output oscillations for natural outputs $q(t), p(t)$, it is also proportional to γ^2. Therefore, it seems natural to introduce the *normalized excitability index* (EI) for the system (4.1.7) as the following quantity:

$$E(\gamma) = \frac{1}{\gamma}\sqrt{\chi(\gamma)}, \qquad (4.1.5)$$

where $\chi(\gamma)$ is the optimum value of the problem (4.1.4).

It is clear that for linear asymptotically stable systems $E(\gamma) = \mathrm{const}$. For nonlinear systems, $E(\gamma)$ is a function of γ that characterizes excitability properties of the nonlinear system. It was introduced in [142] for $\chi(\gamma)$ as in (4.1.4), i.e., for the energy output of the system. However, it can be defined in more general cases if $H_0(x)$ in (4.1.4) is replaced by any other output function of the system. For systems with several inputs and several outputs the excitability indices $E_{ij}(\gamma)$ are introduced in a similar way for every pair of input u_i and output y_j. The concept of EI is related to the concept of *input–output gain*, popular in the modern control theory. Input–output gain is defined as the norm of the operator transforming input function into output function for a dynamical system [227, 414]. If the input–output gain exists, it provides an upper bound for EI. Conversely, if EI is finite, it estimates the minimal value of input–output gain. However, EI cannot be reduced to the input–output gain, since the gain provides the maximum achievable value for upper bound of output, while EI evaluates maximum achievable value for the asymptotic lower bound.

The solution to the problem (4.1.4) for nonlinear systems is quite complicated in most cases. However, we can use approximately a locally optimal or speed-gradient solution

$$u(x) = \gamma \,\mathrm{sign}\,[H_0, H_1], \qquad (4.1.6)$$

where $g(x) = \left.\frac{\partial F(x,u)}{\partial u}\right|_{u=0}$, obtained by maximizing the principal part of $\dot{H}_0$ – the instant growth rate of H_0. It follows from the results of [106] that for small γ the value of $|y(t)|$ achievable with input (4.1.6) for sufficiently large $t \geq 0$ differs from the optimal value $\chi(\gamma)$ by the amount of order γ^2. An important consequence is that excitability index can be estimated directly by applying input (4.1.6) to the system. For real world systems it can be done experimentally. Otherwise, if a system model is available, computer simulations can be used.

4.1.2 Properties of excitability index

Since the excitability index of a system characterizes its sensitivity to a feedback excitation, it is important to relate excitability to inherent dynamical properties of a system. Such bounds for a more general class of strictly passive systems were established in [147]. We present here a slightly modified result.

Consider a system described by state–space equations

$$\dot{x} = F(x, u),\; y = h(x) \qquad (4.1.7)$$

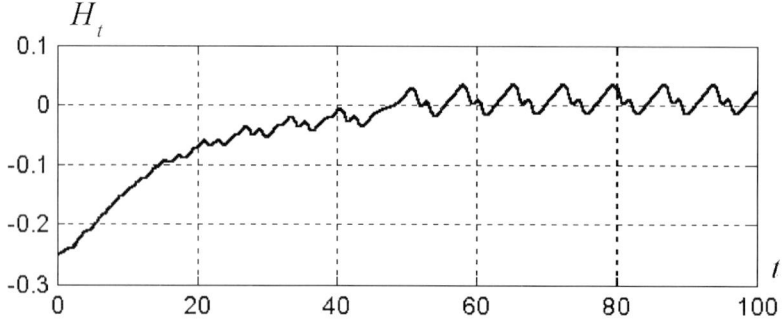

Fig. 4.1.1. Energy dynamics of the damped pendulum $J\ddot\varphi + \varrho\dot\varphi + mgl\sin\varphi = u$ under control $u = \gamma\,\text{sign}\,\dot\varphi$.

where $x \in R^n$ is state vector, u, y are scalar input and output, respectively. Recall that the system (4.1.7) is called *strictly passive with dissipation rate* $\rho(x) \geq 0$ if there exists continuous nonnegative function $V(x) \geq 0$ (storage function) such that for all $t \geq 0$ and any solution $x(t)$ of the system (4.1.7) the following identity holds

$$V(x(t)) = V(x(0)) + \int_0^t (y(s)^\mathrm{T} u(s) - \varrho(x(s)))ds. \tag{4.1.8}$$

The storage function $V(x)$ is an analog of energy for the systems of general form (4.1.7), i.e., identity (4.1.8) can be interpreted as the generalized energy balance equation. Concepts of passivity and dissipativity are widely used in modern nonlinear control theory [324–326, 390, 448].

Computational experiments show that the limit for $t \to \infty$ may not exist even for simple systems, see Fig. 4.1.1, where the process of evolution of the damped pendulum under excitation $u = \gamma\,\text{sign}\,\dot\varphi$ is shown. Therefore, one needs to consider upper and lower limits simultaneously. Introduce the following definition.

Definition 4.1. Let the set of admissible control consist of bounded functions $u(t), 0 \leq t < \infty$ such that the corresponding trajectories $x(t)$ are bounded. Define upper and lower excitability indices $\chi_V^+(\gamma), \chi_V^-(\gamma), 0 \leq \gamma < \infty$ of the system (4.1.7) with respect to the output $V(x)$ as follows:

$$\chi_V^+(\gamma) = \overline{\lim_{t\to\infty}} \sup_{\substack{|u(\cdot)|\leq\gamma\\x(0)=0}} V(x(t)), \tag{4.1.9}$$

$$\chi_V^-(\gamma) = \underline{\lim_{t\to\infty}} \sup_{\substack{|u(\cdot)|\leq\gamma\\x(0)=0}} V(x(t)). \tag{4.1.10}$$

□

4.1 Excitability analysis of dissipative systems

The normalized upper and lower excitability indices $E_V^+(\gamma)$, $E_V^-(\gamma)$ are introduced as follows

$$E^{\pm}(\gamma) = \frac{1}{\gamma}\sqrt{\chi^{\pm}(\gamma)}, \qquad (4.1.11)$$

where

$$\chi^+(\gamma) = \varlimsup_{\substack{t\to\infty \\ |u(\cdot)|\le\gamma \\ x(0)=0}} \mathcal{H}(q(t),\dot{q}(t)). \qquad (4.1.12)$$

To define the excitability indices $\chi_y^+(\gamma)$, $\chi_y^-(\gamma)$ with respect to any output $y = h(x)$ the function $V(x)$ in (4.1.9), (4.1.10) should be replaced by $h(x)$. In the case when the input is vector, $u = \mathrm{col}\{u_1,\ldots,u_m\}$ the value of input is also specified as the vector of maximum values of components $\gamma = \{\gamma_1,\ldots,\gamma_m\}$, where $\gamma_i = \sup_t |u_i(t)|$ and excitability indices are defined as multi-indices. In a more general case of the system with m inputs and l outputs the excitability indices $\chi_y^+(\gamma)$, $\chi_y^-(\gamma)$ are $l \times m$-matrix functions, depending on m arguments. Their (i,j) elements $\chi_{y,i,j}^+(\gamma)$, $\chi_{y,i,j}^-(\gamma)$ are excitability indices "from the input u_i to the input y_j." Note that technical assumption of boundedness of $x(t)$ is not necessary and can be weakened. The main result of this section is the following statement.

Theorem 4.1. *Let the system (4.1.7) be strictly passive with the storage function $V(x)$ and dissipation rate $\varrho(x)$ satisfying inequalities*

$$\alpha_0|y|^2 \le V(x) \le \alpha_1|y|^2 + d, \qquad (4.1.13)$$

$$\varrho_0|y|^2 \le \varrho(x) \le \varrho_1|y|^2 \qquad (4.1.14)$$

for some positive $\alpha_0, \alpha_1, \varrho_0, \varrho_1, d$. Let the set

$$\Omega^- = \left\{ x : h(x) = 0,\ V(x) < \alpha_0\left(\frac{\gamma}{\varrho_1}\right)^2 \right\}$$

not contain whole trajectories of the free system $\dot{x} = F(x,0)$.

Then excitability indices $\chi_V^+(\gamma)$, $\chi_V^-(\gamma)$ with respect to $V(x)$ satisfy inequalities

$$\alpha_0\left(\frac{\gamma}{\varrho_1}\right)^2 \le \chi_V^-(\gamma) \le \chi_V^+(\gamma) \le m\alpha_1\left(\frac{\gamma}{\varrho_0}\right)^2 + d, \qquad (4.1.15)$$

In addition, the lower bound is realized for the speed-gradient control

$$u(t) = \gamma\,\mathrm{sign}\ y(t). \qquad (4.1.16)$$

We see that the action (4.1.6) creates a sort of resonance mode in a nonlinear system: for weakly damped systems even a small action having form (4.1.6) leads to large oscillations of the output and can insert a substantial amount of energy into the system.

4.1.3 Case of Euler–Lagrange systems

The above bounds for excitability indices are expressed in terms of the ratio "(excitation amplitude)/(dissipation)." It is possible to make them more explicit for a class of physical systems described by Euler–Lagrange equations with dissipative forces, if the output is specified as the energy of the system.

Consider an Euler–Lagrange system with dissipation

$$\frac{d}{dt}(A(q)\dot{q}) + R(\dot{q}) + \nabla \Pi(q) = u, \qquad (4.1.17)$$

where $q \in \mathbb{R}^m$ is vector of generalized coordinates, $u = u(t) \in \mathbb{R}^m$ is vector of controlling forces (torques), $A(q)$ is matrix of kinetic energy, $\Pi(q)$ is potential energy, $R(\dot{q})$ is vector of dissipative forces. Let $\mathcal{H}(q,\dot{q})$ be total energy of the system:

$$\mathcal{H}(q,\dot{q}) = \frac{1}{2}\dot{q}^T A(q)\dot{q} + \Pi(q). \qquad (4.1.18)$$

Normalized upper and lower excitability indices $E^+(\gamma)$, $E^-(\gamma)$ are introduced as follows

$$E^{\pm}(\gamma) = \frac{1}{\gamma}\sqrt{\chi^{\pm}(\gamma)}, \qquad (4.1.19)$$

where

$$\chi^+(\gamma) = \overline{\lim_{t\to\infty}} \sup_{\substack{|u(\cdot)|\leq\gamma \\ x(0)=0}} \mathcal{H}(q(t),\dot{q}(t)), \qquad (4.1.20)$$

$$\chi^-(\gamma) = \underline{\lim_{t\to\infty}} \sup_{\substack{|u(\cdot)|\leq\gamma \\ x(0)=0}} \mathcal{H}(q(t),\dot{q}(t)), \qquad (4.1.21)$$

Theorem 4.2. *Let*

$$0 < \alpha^- \leq \lambda_i(A(q)) \leq \alpha^+,$$
$$\varrho^-|\dot{q}|^2 \leq R(\dot{q})^T q \leq \varrho^+|\dot{q}|^2,$$
$$0 \leq \Pi(q) \leq d.$$

Then

$$\frac{\alpha^-}{2}\left(\frac{\gamma}{\varrho^+}\right)^2 \leq \chi^-(\gamma) \leq \chi^+(\gamma) \leq m\alpha^+\left(\frac{\gamma}{\varrho^-}\right)^2 + d. \qquad (4.1.22)$$

Corollary 4.1. *If* $R(\dot{q}) = \varrho\dot{q}$ *and* $\varrho \to 0$, *then*

$$E^{\pm}(\gamma) \sim \frac{C_{\pm}}{\varrho}. \qquad (4.1.23)$$

Remark 4.1. Locally optimal control is

$$u = \gamma \, sign(\dot{q}). \qquad (4.1.24)$$

It will be suboptimal for small $\gamma > 0$. Action (4.1.24) creates a resonance regime of the system (4.1.17) because the order of the estimate (4.1.23) coincides with the order of magnitude frequency response for linear systems. Relations (4.1.22) or (4.1.23) can be considered as transformation of energy laws for dissipative Euler–Lagrange systems which can be loosely formulated as follows:

> The level of energy achievable by means of control of the power γ for controllable Hamiltonian or Lagrangian system with small dissipation of the degree ρ has the order $(\gamma/\rho)^2$

4.1.4 Example: Excitation of the dumped pendulum

For systems with one degree of freedom the previous results simplify.

Again consider the example of the Chapter 3 taking into account the linear viscous friction with friction coefficient ϱ. It follows from Theorem 4.2 that

$$0.5\left(\frac{\gamma}{\varrho}\right)^2 \leq \overline{H}_0 \leq \left(\frac{\gamma}{\varrho}\right)^2 + 2\omega_0^2, \qquad (4.1.25)$$

where $\overline{H}_0$ is the limit value of achievable energy.

The control law realizing the estimate (4.1.25) takes the form

$$u = \gamma \operatorname{sign}(\dot{\varphi}). \qquad (4.1.26)$$

More accurate estimates can be obtained under additional assumptions. For example, assume that the steady oscillation mode in the closed-loop system is close to harmonic one (such an assumption holds for small γ). In the steady mode the energy lost due to dissipation during the period of oscillation is equal to the energy supplied by control. Evaluation of the energies as integrals over the period yields the following estimate of the balanced energy

$$\overline{H} \approx \frac{8}{\pi^2}\left(\frac{\gamma}{\varrho}\right)^2, \qquad (4.1.27)$$

The estimate (4.1.27) agrees with (4.1.25), since $0.5 < 8/\pi^2 < 1$. Note that in the above derivation no assumption concerning the shape of the potential is needed. Therefore approximate bound (4.1.27) can be used for oscillators with different shape of nonlinearity.

The obtained estimates allow to evaluate the excitability degree and resonance properties of nonlinear systems and to obtain an additional information about their dynamical characteristics.

4.1.5 Example: Excitation of the Duffing system

Note that for the derivation of the estimates (4.1.15) the boundedness of potential (assumption (4.1.13)) is essential. However, a careful choice of the storage function in Theorem 4.1 sometimes may help. For example, consider the Duffing system

$$\ddot{\varphi} + \varrho\dot{\varphi} - a\varphi + b\varphi^3 = u(t), \qquad (4.1.28)$$

where $a > 0$, $b > 0$, $\varrho > 0$. The Duffing system describes dynamics of buckled beams, plates, magnetic devices, and many other physical processes. It has become one of paradigmatic models in the area of nonlinear dynamics. The model (4.1.28) belongs to the class of Euler–Lagrange systems with dissipation and can be represented as the system of two differential equations

$$\dot{\varphi} = y, \dot{y} = -\varrho y + \Pi'(\varphi) + u(t), \qquad (4.1.29)$$

where potential $\Pi(\varphi)$ has the form

$$\Pi(\varphi) = \frac{b}{4}\varphi^4 - \frac{a}{2}\varphi^2. \qquad (4.1.30)$$

The potential (4.1.30) has two wells with minima in the points $\varphi_{1,2} = \pm\sqrt{a/b}$ and the saddle point $\varphi = 0$. The energy function of the free Duffing system is

$$\frac{1}{2}\dot{\varphi}^2 + \Pi(\varphi) \qquad (4.1.31)$$

and for its excitation the speed-gradient algorithm

$$u(t) = \gamma\mathrm{sign}\dot{\varphi} \qquad (4.1.32)$$

can be used. Though the potential (4.1.30) is unbounded and Theorem 4.1 does not apply, the left inequality in (4.1.22) is still fulfilled and the following lower bound for normalized excitability index of Duffing system holds:

$$E^-(\gamma) \geq \frac{1}{\varrho\sqrt{2}}.$$

To obtain an upper bound the modified storage function

$$V(\varphi, \dot{\varphi}) = \dot{\varphi}^2/2 + \Pi(\varphi) + \varepsilon\varphi\dot{\varphi} \qquad (4.1.33)$$

for small ε can be used. With the function (4.1.33) it is possible to prove boundedness of the system solution under bounded control. Consider a more general class of oscillators (4.1.29) with potentials satisfying inequalities

$$\Pi(\varphi) \geq c_0|\varphi|^2 - d_0, \quad \varphi\Pi'(\varphi) \geq c_1\Pi(\varphi) - d_1. \qquad (4.1.34)$$

Obviously, the Duffing potential (4.1.30) satisfies (4.1.34) with $c_0 = a/2$, $d_0 = 2a^2/b$, $c_1 = 4$, $d_1 = a^2/b$. It can be shown that for oscillators (4.1.29) with potentials satisfying inequalities (4.1.34) the upper bound for excitability index is as follows:

$$\chi^+(\gamma) \leq R_1\left(\frac{\gamma}{\varrho}\right)^2 + R_0, \qquad (4.1.35)$$

where $R_1 = \max\{2c_1, c_1^2\}$, $R_0 > 0$. Particularly, for Duffing system $R_1 = 16$ and the bound for normalized excitability index for small ρ is

$$E^+(\gamma) \le \frac{4}{\varrho}.$$

More tight estimate for excitability index can be obtained using harmonic balance of energy. Since in the derivation of (4.1.27) no assumption concerning the shape of the potential is used, the following approximation holds

$$E^\pm(\gamma) \approx \frac{2\sqrt{2}}{\pi \varrho} \approx 0.9/\varrho.$$

4.2 Feedback resonance

The concept of *resonance* has numerous applications in physics and mechanics. In essence it is that small resonant force applied to a system leads to significant changes in system behavior. First clear description of resonance phenomenon was given by Galileo Galilei in "Discorsi a Dimostrazioni Matemaci," published in 1638 (see [405]):

> ...Pendulum at rest although very heavy, can be put into motion, and very significant if we stop our breath when it is coming back and blow again at the instant, corresponding to its swing.

The resonance phenomenon is well understood in its application to linear systems. However, if the dynamics of the system is nonlinear, the resonance is much more complicated because interaction of different harmonic signals in nonlinear system may create complex and even chaotic behavior [77, 253, 385]. The reason is, that the natural frequency of a nonlinear system depends on the amplitude of oscillations.

In [140, 142] the idea to create resonance in a nonlinear oscillator by changing the frequency of external action as a function of oscillation amplitude was pursued. Consider the controlled 1-DOF oscillator modeled after appropriate rescaling by the differential equation

$$\ddot{\varphi} + \Pi(\varphi)' = u, \qquad (4.2.36)$$

where φ is the phase coordinate, $\Pi(\varphi)$ is potential energy function, u is controlling variable. The state vector of the system (4.2.36) is $x = (\varphi, \dot{\varphi})$ and its important characteristics is the total energy $H(\varphi, \dot{\varphi}) = \frac{1}{2}\dot{\varphi}^2 + \Pi(\varphi)$. The state vector of the uncontrolled (free) system moves along the energy surface (curve) $H(\varphi, \dot{\varphi}) = H_0$. The behavior of the free system depends on the shape of $\Pi(\varphi)$ and the value of H_0. For example, for simple pendulum we have $\Pi(\varphi) = \omega_0^2(1 - \cos\varphi) \ge 0$. Obviously, choosing $H_0 : 0 < H_0 < 2\omega_0^2$ we obtain oscillatory motion with amplitude $\varphi_0 = \arccos(1 - H_0/\omega_0^2)$. For

$H_0 = 2\omega_0^2$ the motion along the separatrix including upper equilibrium is observed, while for $H_0 > 2\omega_0^2$ the energy curves become infinite and the system exhibits permanent rotation with average angular velocity $<\dot\varphi> \approx \sqrt{2H_0}$.

Let us ask, whether it possible to significantly change the energy (i.e., behavior) of the system by means of arbitrarily small controlling action.

The answer is well known when the potential is quadratic, $\Pi(\varphi) = \frac{1}{2}\omega_0^2\varphi^2$, i.e., system dynamics are linear:

$$\ddot\varphi + \omega_0^2 \varphi = u. \tag{4.2.37}$$

In this case we may use harmonic external action

$$u(t) = \bar u \sin \omega t \tag{4.2.38}$$

and for $\omega = \omega_0$ watch the unbounded resonance solution $\varphi(t) = -\bar u t / 2\omega_0 \cos\omega_0 t$.

However for nonlinear oscillators the resonant motions are more complicated with interchange of energy absorption and emission. It is well known that even for simple pendulum the harmonic excitation can create chaotic motions. The reason is, that the natural frequency of a nonlinear system depends on the amplitude of oscillations.

Therefore the question is: how to create a fully resonance mode in a nonlinear oscillator? One possible answer is: the frequency of external action should be changing as a function of oscillation amplitude. In order to implement such a solution one may choose $u(t)$ as a function of the current measurements $\varphi(t), \dot\varphi(t)$ which exactly means introducing a feedback

$$u(t) = U(\varphi(t), \dot\varphi(t)). \tag{4.2.39}$$

The problem can be reformulated as follows: find the feedback law (4.2.39) allowing to achieve the given energy surface $H(\varphi, \dot\varphi) = H_*$. Such a problem falls into the field of control theory. To solve it we suggest to use speed-gradient method, see Section 2.4.2. For the system (4.2.36) the speed-gradient method with the choice of the goal function $Q(x) = [H(x) - H_*]^2$ produces simple feedback laws:

$$u = -\gamma(H - H_*)\dot\varphi, \tag{4.2.40}$$

$$u = -\gamma \operatorname{sign}(H - H_*) \cdot \operatorname{sign}\dot\varphi, \tag{4.2.41}$$

where $\gamma > 0$, $\operatorname{sign}(H) = 1$, for $H > 0$, $\operatorname{sign}(H) = -1$ for $H < 0$ and $\operatorname{sign}(0) = 0$. It follows from Theorem 3.1 that the goal $H(x(t)) \to H_*$ in the system (4.2.36), (4.2.40) (or (4.2.36), (4.2.41)) will be achieved from almost all initial conditions provided that the potential $\Pi(\varphi)$ is smooth and its stationary points are isolated. It is worth noting that since the motion of the controlled system belongs to the finite energy layer between H_0 and H_*, the right-hand side of (4.2.40) is bounded. Therefore, choosing sufficiently small gain γ we can achieve the given energy surface $H = H_*$ by means of *arbitrarily small*

4.2 Feedback resonance

control. Of course this seemingly surprising result holds only for conservative (lossless) systems.

Let now losses be taken into account, i.e., system is modeled as

$$\ddot{\varphi} + \varrho\dot{\varphi} + \Pi(\varphi)' = u, \qquad (4.2.42)$$

where $\varrho > 0$ is the damping coefficient. Then it is not possible any more to reach an arbitrary level of energy. The lower bound $\overline{H}$ of the energy value reachable by a feedback of amplitude $\overline{u}$ can be calculated using Theorem 4.1 as

$$\overline{H} = \frac{1}{2}\left(\frac{\overline{u}}{\varrho}\right)^2. \qquad (4.2.43)$$

In order to achieve the energy (4.2.43) the parameters of feedback should be chosen properly. Namely, parameter values of the algorithm (4.2.40) providing energy (4.2.43) under restriction $|u(t)| \leq \overline{u}$ are as follows: $H_* = 3\overline{H}, \gamma = \varrho/(2\overline{H})$. For the algorithm (4.2.41) with $\gamma = \overline{u}$ any value H_* exceeding $\overline{H}$ is appropriate as follows from Theorem 4.1.

Note that H_* does not have the meaning of the desired energy level in presence of losses. It leaves some freedom of parameter choice. Exploiting this observation we may take H_* sufficiently large in the algorithm (4.2.41) and arrive to its simplified form

$$u = -\gamma \operatorname{sign}\dot{\varphi}, \qquad (4.2.44)$$

that looks like introducing negative Coulomb friction into the system.

It is worth to compare the bound (4.2.43) with the energy level achievable for linear oscillator

$$\ddot{\varphi} + \varrho\dot{\varphi} + \omega_0^2\varphi = u(t), \qquad (4.2.45)$$

where $\varrho > 0$ is the damping coefficient, by harmonic (nonfeedback) action. The response of the model to the harmonics $u(t) = \overline{u}\sin\omega t$ is also harmonics $\varphi(t) = A\sin(\omega t + \varphi_0)$ with the amplitude

$$A = \frac{\overline{u}}{\sqrt{(\omega^2 - \omega_0^2)^2 + \varrho^2\omega^2}}. \qquad (4.2.46)$$

Let ϱ be small, $\varrho^2 < 2\omega_0^2$. Then A reaches its maximum for resonant frequency: $\omega^2 = \omega_0^2 - \varrho^2/4$, and the system energy averaged over the period is

$$\overline{H} = \frac{1}{2}\left(\frac{\overline{u}}{\varrho}\right)^2 + O(\varrho^2), \qquad (4.2.47)$$

A comparison of (4.2.43) and (4.2.47) shows that for a nonlinear oscillator affected by feedback, the change of energy can reach the limit achievable for linear oscillator by harmonic (nonfeedback) action, at least in the case of small damping. Therefore, feedback allows a nonlinear oscillator to achieve as deep a resonance as can be achieved by harmonic excitation for the linear case.

Since such a mode is not achievable without feedback for nonlinear systems, the phenomenon can be called *feedback resonance* (f-*resonance*) [140]. It should be noted that the understanding of resonance in physics has remained practically the same since the time of Galileo. In most works on the subject, the input action is assumed to be harmonic (periodic, at the most). In the classical book by Andronov et al. [31], whose first edition was published in 1937, the authors introduce the concept of autoresonance as "resonance generated by a force caused by the motion of the system," i.e., the possibility of actions in a feedback form was pointed to. However, in [31] the authors only considered the case of a linear system of the second order with a relay in the feedback loop. For that case the existence of limit cycles and estimates of their size were established. The system was assumed to be closed, i.e., what was really studied was the *internal* resonance in the system. It probably explains the origin of the term "autoresonance."

Autoresonance modes were also analyzed in [47, 48]. Particularly, it was proposed in [47] to use SG-algorithms for the tuning of a nonlinear system parameters in order to keep its resonance mode. In a number of papers [35, 130, 274, 294] the term "autoresonance" was used in the case of the external excitation close to a periodic function, with slowly ("adiabatically") varying frequency. On the contrary, the feedback resonance phenomenon occurs under external excitation that changes its spectrum in the course of the process.

It is interesting that Galileo's description of resonance in fact admits a possibility of feedback. Even more, it suggests how feedback can be used to force the pendulum into the resonant mode: one simply needs to blow "stop our breath when it is coming back and blow again at the instant, corresponding to its swing."

4.3 Excitability index of pendulum systems

The excitability index allows to measure the stabilizing properties of a system and its closeness to the stability margin. In that sense the excitability index may be used as a substitute of magnitude frequency response in absolute stability criteria for nonlinear systems [142, 143].

Excitability index can be measured in the course of simulation or experimental study of the system, like standard frequency-domain response is used by many control engineers. However, there is a significant difference. In order to evaluate frequency-domain response the input of the system is affected with the harmonic excitation signal (4.2.38) having constant magnitude and variable frequency. In order to evaluate excitability index the magnitude (level) of the input signal is variable, while the signal itself is designed as a feedback and does not have any specific frequency.

Exact evaluation of the indices $E^{\pm}(\gamma)$ is complicated because of involved optimal synthesis required. However, to obtain their lower bounds, the output of the system affected with the speed-gradient signal (4.1.16) can be measured.

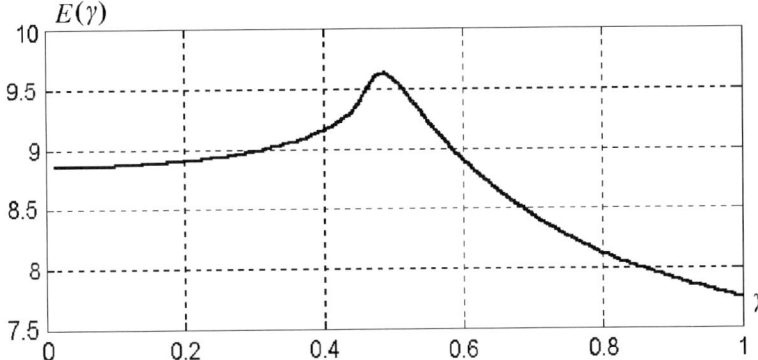

Fig. 4.3.2. Excitability index of pendulum (4.3.48) for $\varrho = 0.1\ c^{-1}$, $\omega_0^2 = 10\ c^{-2}$ (simulation).

Such a signal is locally optimal and, similarly to [106] it can be shown that for small γ the action (4.1.16) provides an approximate value for $E(\gamma)$ with accuracy of the order γ.

Below are a few examples of the excitability index graphs for pendulum systems. First, evaluate the excitability index of simple pendulum with the model

$$\ddot{\varphi}(t) + \varrho\dot{\varphi}(t) + \omega_0^2 \sin\varphi(t) = u(t) \qquad (4.3.48)$$

with respect to energy output for locally optimal speed-gradient action

$$u = \gamma\ \mathrm{sign}\ \dot{\varphi}. \qquad (4.3.49)$$

The results for parameter values $\omega_0^2 = 10.0\ s^{-2}$, $\varrho = 0.1\ s^{-1}$, and initial conditions $\varphi(0) = 0$, $\dot{\varphi}(0) = 10^{-10}\ s^{-1}$ are shown in Fig. 4.3.2. It is interesting to note that the graph of Fig. 4.3.2 agrees reasonably with the results of the Example 4.1, where $E(\gamma) \to \frac{20}{\pi}\sqrt{2}J^{1/2} \approx 9.0\ J^{1/2}$ when $\gamma \to 0$. In Fig. 4.3.3 the experimental results of excitability index measurements are presented. The experiments were carried out on the mechatronic pendulum system developed in the Institute for Problems of Mechanical Engineering in St. Petersburg [111]. It is seen that simulation and experimental results qualitatively coincide. The peak of the graph $E(\gamma)$ corresponds to the level of input signal swinging the pendulum to the upright position. It means that observing the curve $E(\gamma)$ one can estimate critical levels of the system potential energy. A typical plot of the system output $\dot{\varphi}(t)$ is shown in Fig. 4.3.4.

The excitability plots for complex systems may look very intricate. It contains information about delicate dynamical properties of a system but it may not be easy to extract such an information. Consider, for example, the system of two-coupled pendulums

$$\begin{aligned}\ddot{\varphi}_1 + \varrho_1\dot{\varphi}_1 + \omega_1^2\sin\varphi_1 + k(\varphi_1 - \varphi_2) &= u(t),\\ \ddot{\varphi}_2 + \varrho_2\dot{\varphi}_2 + \omega_2^2\sin\varphi_2 + k(\varphi_2 - \varphi_1) &= 0,\end{aligned} \qquad (4.3.50)$$

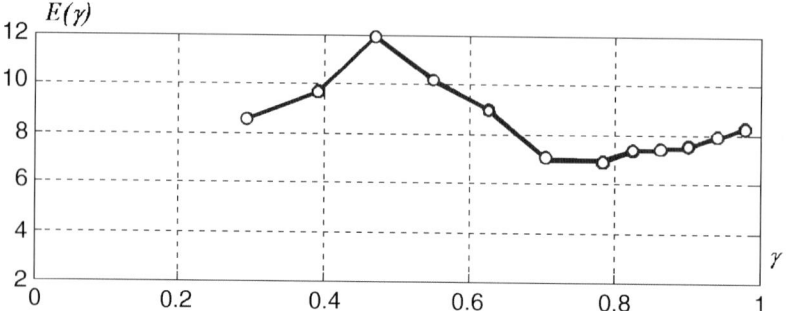

Fig. 4.3.3. Excitability index of pendulum (experiment).

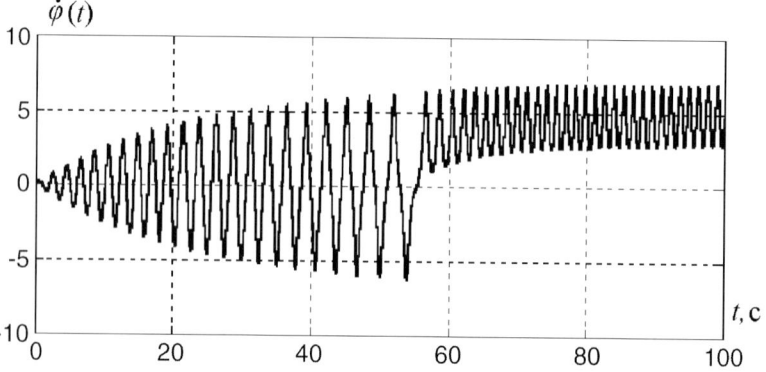

Fig. 4.3.4. Typical trajectory of the controlled pendulum (4.3.48), (4.3.49) for $\gamma = 1.25$.

where $k > 0$ is coupling coefficient. Let the system output be its energy

$$H = \frac{1}{2}\left(\dot\varphi_1^2 + \dot\varphi_2^2\right) + \omega_1^2\left(1 - \cos\varphi_1\right) + \omega_2^2\left(1 - \cos\varphi_2\right) + \frac{k}{2}\left(\varphi_1^2 + \varphi_2^2\right),$$

and let the system be excited with the action similar to (4.3.49):

$$u = \gamma \operatorname{sign} \dot\varphi_1. \tag{4.3.51}$$

Let us start from the case $k = 0$ and $\varphi_2(0) = \dot\varphi_2(0) = 0$. Then the excitability plot $E(\gamma)$ coincides with the excitability plot of a single pendulum, see Fig. 4.3.2. As the coupling k grows, the dynamics of the nonlinear system (4.3.50) becomes more complex which is seen from the graph of $E(\gamma)$. The excitability plots for $k = 0.25,\ 0.5,\ 1.0,\ 5.0$ are shown in Figs. 4.3.5, 4.3.7, 4.3.9, 4.3.11. In Fig. 4.3.6, 4.3.8, 4.3.10, 4.3.12 the graphs of typical processes $\dot\varphi_1(t)$, $\dot\varphi_2(t)$. are presented. Initial conditions for simulation and parameter values were chosen as follows: $\varphi_1(0) = \varphi_2(0) = 0$, $\dot\varphi_1(0) = 10^{-10}\ s^{-1}$, $\dot\varphi_2(0) = 0\ s^{-1}$, $\varrho_1 = \varrho_2 = 0.1\ s^{-1}$, $\omega_1^2 = \omega_2^2 = 10\ s^{-2}$. Simulation time for evaluation of $E(\gamma)$ was taken 500 s. For averaging the moving average method was used with the window of 500 samples and sampling interval 0.05 c.

4.3 Excitability index of pendulum systems   63

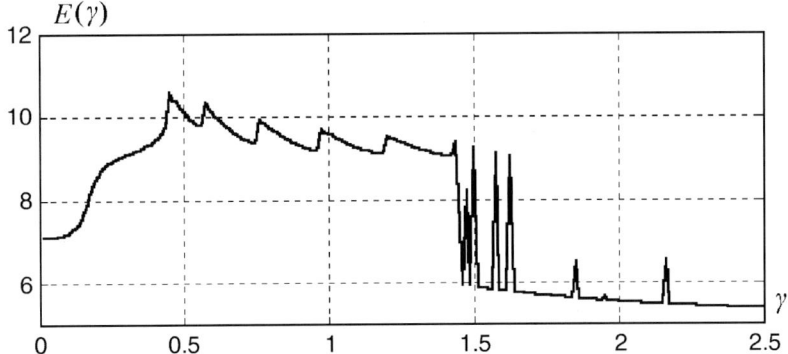

Fig. 4.3.5. Excitability index of the two-pendulum system (4.3.50) for $k = 0.25$.

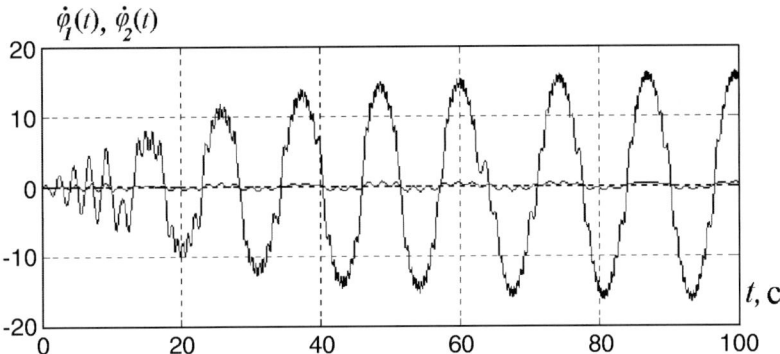

Fig. 4.3.6. Typical trajectory of the system (4.3.50), (4.3.51) for $k = 0.25$, $\gamma = 1.25$.

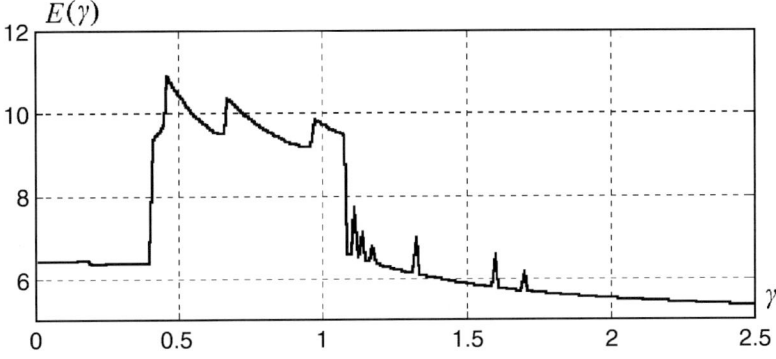

Fig. 4.3.7. Excitability index of the two-pendulum system (4.3.50) for $k = 0.5$.

64   4 Control of Dissipative Systems

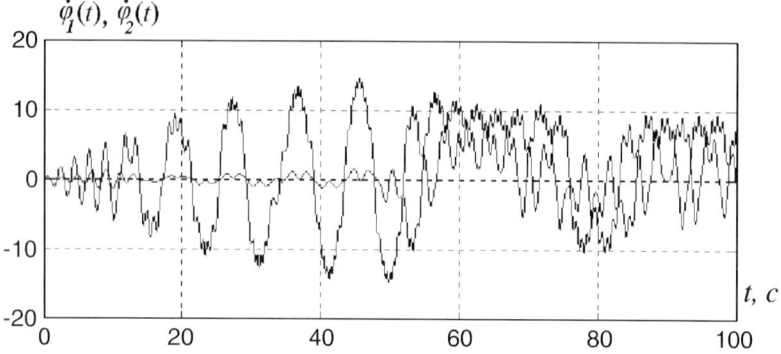

Fig. 4.3.8. Typical trajectory of the system (4.3.50), (4.3.51) for $k = 0.5$, $\gamma = 1.25$.

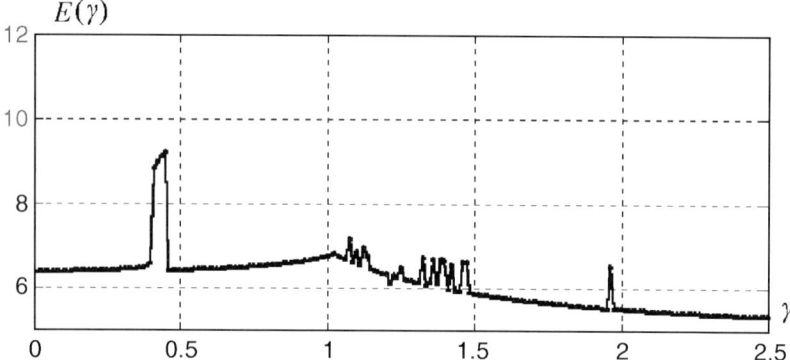

Fig. 4.3.9. Excitability index of the two-pendulum system (4.3.50) for $k = 1.0$.

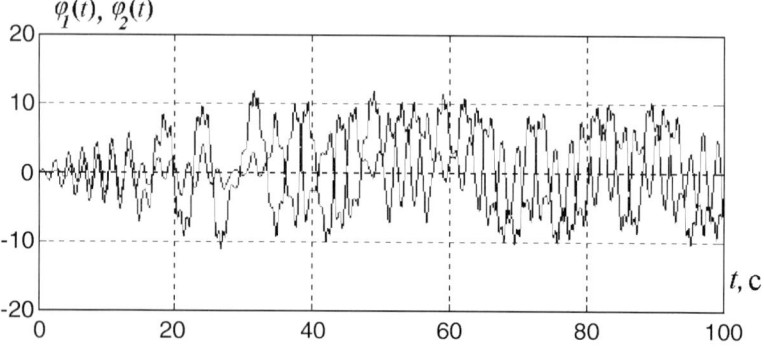

Fig. 4.3.10. Typical trajectory of the system (4.3.50), (4.3.51) for $k = 1.0$, $\gamma = 1.25$.

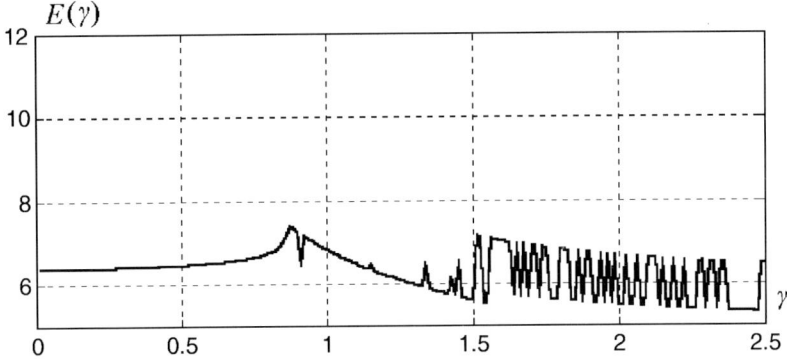

Fig. 4.3.11. Excitability index of the two-pendulum system (4.3.50) for $k = 5.0$.

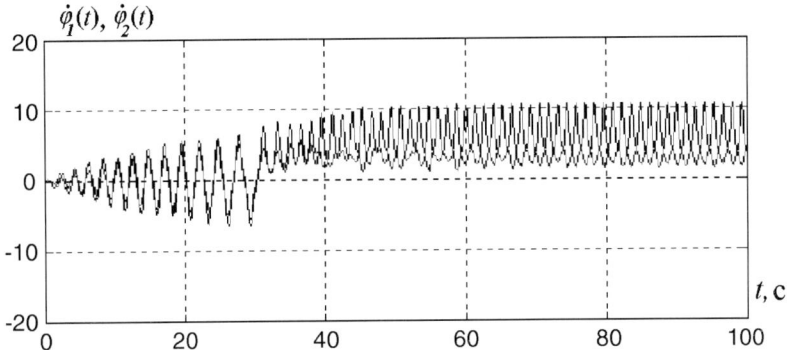

Fig. 4.3.12. Typical trajectory of the system (4.3.50), (4.3.51) for $k = 5.0, \gamma = 1.25$.

It is seen from the pictures that growth of the coupling coefficient k implies that the excitability plot $E(\gamma)$ becomes more and more twisted and jagged and the number of its extremums increases. Extremums of $E(\gamma)$ correspond to the *bifurcations* (qualitative change of the system trajectories), caused by changes of the number of the pendulum swings before a rotatory motion occurs and changes of the number of pendulum relative slippings. For γ corresponding to the multiextremum plots $E(\gamma)$ the motion of pendulums looks irregular and even chaotic.

Excitability plots with respect to other output provide additional information about the system dynamics. In Fig. 4.3.13 and Fig. 4.3.14 the excitability plots of the system (4.3.48), (4.3.50) with respect to "velocity" output $\dot{\varphi}^2$ and with respect to $\dot{\varphi}_1^2 + \dot{\varphi}_2^2$ (kinetic energy), respectively are shown. Deep holes correspond to predominance of trajectories, "freezing" near upper (unstable) equilibrium.

66   4 Control of Dissipative Systems

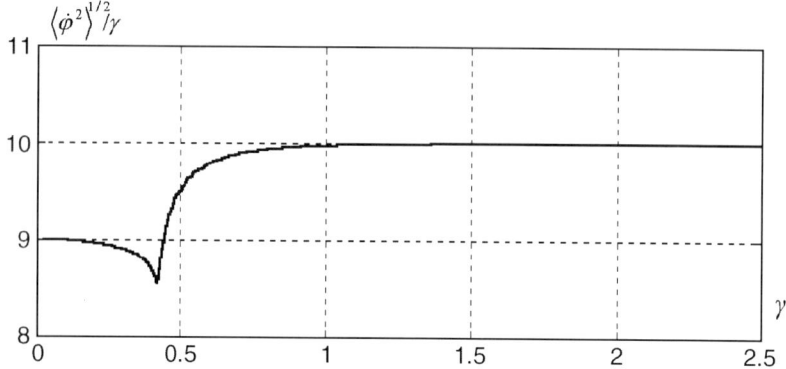

Fig. 4.3.13. Excitability index of the pendulum (4.3.48) with respect to the output $\dot\varphi^2$ for $k = 1.0$.

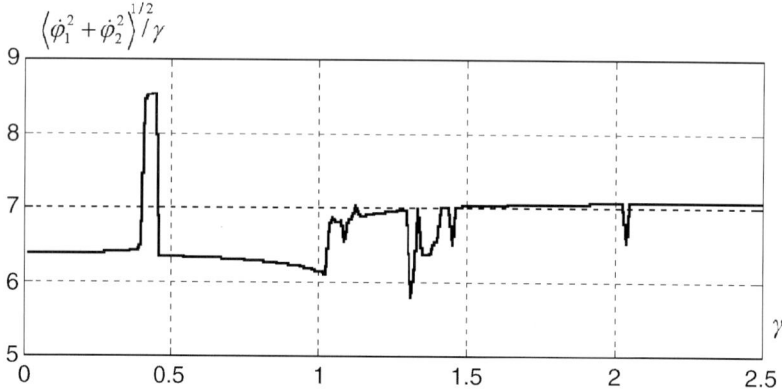

Fig. 4.3.14. Excitability index of the two-pendulum system (4.3.48), (4.3.49) with respect to the output $\dot\varphi_1^2 + \dot\varphi_2^2$ for $k = 1.0$.

To conclude, the excitability index plays the same role of a measure of resonant properties for nonlinear systems, as the magnitude frequency response plays for linear systems. Evaluation of its analytic bounds can provide a preliminary estimate. More accurate estimates can be obtained via numerical or physical experiments. Such an approach may be helpful for study of complex oscillatory processes.

5
Controlled Synchronization

In this chapter, general definitions of synchronization notion covering both controlled and uncontrolled synchronization are given. Different types of synchronization are exemplified and discussed. Several approaches to controlled synchronization systems design are described: observer-based synchronization including Pecora–Carroll scheme, passification-based adaptive synchronization, and speed-gradient synchronization of two-pendulum systems.

5.1 Definitions of synchronization

The term *synchronization* in scientific colloquial use means coordination or agreement in time of two or several processes or objects. For example, it may be coincidence or closeness of the observable variables for two or several systems. Synchronization may also manifest itself as correlated in time changes of some quantitative characteristics of the systems.

In some cases the synchronous regime arises due to natural properties of the processes themselves and their natural interaction. A well-known example is *frequency synchronization* of oscillating or rotating bodies. Such a phenomenon is called *self-synchronization*. In other cases, to achieve synchronization one needs to introduce special actions or impose special constraints. Then we will speak about *forced* or *controlled* synchronization understood as the stage in the time history of the system required to achieve a synchronous regime.

The synchronization phenomenon has numerous applications in vibrational technologies [71–73], in electronics and telecommunications [266, 272], and in other fields [350].

Since the middle of the 1980s there had been significant interest in the so-called chaotic synchronization, when the synchronized subsystems continue to perform complex chaotic oscillations even after the synchronous mode is achieved [7, 81, 164, 170, 343, 350]. A broadly discussed topic is the use of chaotic synchronization for improving security and reliability of the information transmission. A number of special issues of the international journals are devoted to those problems [101, 102, 206–209, 287, 422].

Recently an increasing interest is observed in the controlled synchronization where additional actions or feedbacks are used to achieve synchronization mode. Earlier such problems were studied mainly for linear systems [197, 298]. Controlled synchronization allows to extend the class of the systems possessing synchronous modes and increase their stability and robustness.

Some work is being pursued on studying the synchronization of oscillator arrays and lattices, with applications to synchronization of the biological systems, artificial and/or natural neurons, etc. [64, 350, 364].

Due to emerging interest in synchronization from different scientific communities, a paradoxical situation has arisen: some groups of researchers are not aware of results obtained by the researchers working in other directions. Consequently, achievements of one group may not be used by the other ones. Moreover, understandings of the term "synchronization" may differ significantly from one group to another one.

In order to study different problems from a unified viewpoint it seems useful to formulate a general definition of synchronization, encompassing the main existing definitions. Perhaps, the first definition of such kind, covering both controlled and noncontrolled synchronization was proposed in [75] and extended in [76, 111].

In this section a modified definition is given, following [74, 147]. A number of examples demonstrating features of the proposed definition and its relation to the other existing definitions are presented.

5.1.1 Evolution of the synchronization concept

First versions of the general definition for periodic processes were proposed in [71] (coincidence or multiplicity of the average frequencies of oscillatory or rotatory motions) and in [190] (existence of an asymptotically stable invariant torus of the dimension $n-m$, where m is the degree of synchronization). In [71] it was also pointed out that synchronization can be understood as coincidence of some functionals depending on the systems coordinates (for example, time when some coordinate crosses some level, or time when some coordinate takes its extremum value).

Studies of synchronization for chaotic processes led to a number of new versions of the synchronization concept: coordinate (identical) synchronization, [7, 170, 343], generalized synchronization [384], phase synchronization [375], master–slave synchronization [343], and others. According to [7] synchronization is understood as the existence of a homeomorphism $g : \pi_1(A_c) \to \pi_2(A_c)$, such that $g\left(\pi_1(x_1(t))\right) \to \pi_2\left(x_2(t + \alpha(t))\right)$, where $A(c)$ is attractor of the interconnected system, π_1 and π_2 are projectors to the state spaces of the partial processes, $\alpha(t)$ is asymptotically constant phase shift.

The general definition of synchronization covering both controlled and noncontrolled synchronization was proposed in [75] and extended in [76, 111].

Some related definitions were proposed later in [82, 90]. In [219] the definition based upon the concept of the invariant manifold of dynamical system was studied which covers both coordinate and generalized synchronization. Below a modified version of the definition of [75, 76, 111] is described. It allows the reader to easily obtain different existing definitions as its special cases. The definition consists of two levels. The first one deals with observable processes (functions of time) irrespective of the mechanisms (systems) generating them and/or controlling them.

5.1.2 Synchronization of processes

Consider a number of processes (functions of time), the state of each at the time t being described by some vector $x^{(i)}(t)$, $i = 1, 2 \ldots, k$ where $0 \leq t < \infty$. The processes $x^{(i)}$ may represent the states or observable output functions of some systems, but in this section the mechanism generating the processes is inessential. Assume for simplicity that all the processes belong to the same functional space $\mathcal{X}$.

Let a certain characteristic of the processes be defined as the time-dependent family of mappings $C_t : \mathcal{X} \to \mathcal{C}$, where C is the set of possible values of C_t. The characteristic C_t will be called the *synchronization index*. It is important that the index C_t is the same for all the processes. The value of C_t may be a scalar, a vector or a matrix, as well as a function (e.g., spectrum of the process).

Let, finally, the set of vector-functions $F_i : \mathcal{C} \to \mathbb{R}^m$, $i = 1, \ldots, k$ be given. The functions F_i are called *comparison functions*.

Definition 5.1. We will say that the synchronization of the processes $x^{(i)}(t)$, $i = 1, \ldots, k$ with respect to the index C_t and comparison functions F_i occurs (or the processes are synchronized with respect to the index C_t and comparison functions F_i) if there exist real numbers τ_i, $i = 1, \ldots, k$ (called *time shifts or phase shifts*) such that the following relations hold for all t:

$$F_1\left(C_{t+\tau_1}[x_1]\right) = F_2\left(C_{t+\tau_2}[x_2]\right) = \ldots = F_k\left(C_{t+\tau_k}[x_k]\right). \quad (5.1.1)$$

In addition, *approximate synchronization* (ε-synchronization) is understood as approximate fulfillment of the relations (5.1.1), with accuracy of ε:

$$\left\|F_i\left(C_{t+\tau_i}[x_i]\right) - F_j\left(C_{t+\tau_j}[x_j]\right)\right\| \leq \varepsilon \quad \forall i, j, \quad t \geq 0. \quad (5.1.2)$$

Asymptotic synchronization is understood as fulfillment of (5.1.1) when $t \to \infty$:

$$\lim_{t \to \infty} \left\|F_i\left(C_{t+\tau_i}[x_i]\right) - F_j\left(C_{t+\tau_j}[x_j]\right)\right\| = 0. \quad (5.1.3)$$

In a similar way other versions of synchronization notion can be defined. For example, *asymptotic approximate synchronization* is defined as asymptotic fulfillment of the relations (5.1.1) with accuracy of ε:

$$\overline{\lim}_{t\to\infty} \left\| F_i\left(C_{t+\tau_i}[x_i]\right) - F_j\left(C_{t+\tau_j}[x_j]\right) \right\| \le \varepsilon \quad \forall i,j, \quad t \ge 0, \quad (5.1.4)$$

where ε is the asymptotic accuracy threshold.

In the case when some averaging operator $\langle \cdot \rangle_t$ on $0 \le s \le t$ is specified, the *synchronization in the average* can be defined as fulfillment of relations

$$\langle Q_s \rangle_t < \varepsilon, \quad (5.1.5)$$

for all $t \ge 0$, where Q_t is a certain scalar function (*desynchronization measure*), evaluating the deflection from the synchronous mode. Often the averaging operator is specified as an integral operator $\langle Q_s \rangle_t = \frac{1}{t}\int_0^t Q_s ds$, while the desynchronization measure Q_t is defined as mean square deviation from the synchronous mode:

$$Q_t = \sum_{i,j=1}^{k} \left\| F_i\left(C_{t+\tau_i}[x_i]\right) - F_j\left(C_{t+\tau_j}[x_j]\right) \right\|^2. \quad (5.1.6)$$

The possibility of introducing a scalar measure of desynchronization is an important feature of general definition. It opens a way to regular procedures of the control synchronization algorithms design. For example, such a measure can be used as the goal function for gradient design (for discrete time) or speed-gradient design (for continuous time).

Remark 5.1. Sometimes it is convenient to write (5.1.1) in the form

$$F_i\left(C_t[x^{(i)}(t+\tau_i)]\right) - F_k\left(C_t[x^{(k)}(t+\tau_k)]\right) = 0, \quad i=1,2,\ldots,k-1. \quad (5.1.7)$$

The introduced definition encompasses most existing forms of the synchronous behavior of the processes. Let us consider some examples.

5.1.3 Examples

Example 5.1. (Frequency (Huygens) synchronization). This most well known kind of synchronization is defined for the processes possessing well-defined frequencies ω_i, e.g., periodic (oscillatory or rotatory) processes. Introduce the characteristics (synchronization index) C_t as the average velocity over the interval $0 \le s \le t$, i.e., $C_t = \omega_t = \langle \dot{x} \rangle_t$. Since the frequency synchronization criterion is defined by relations $\omega_t = n_i \omega^*$, where for some integer n_i (synchronization multiplicities), ω^* is the so-called *synchronous frequency*, it is natural to introduce comparison functions as $F_i(\omega_t) = \omega_t / n_i$. For the case $n_i = 1, i = 1, \ldots, k$ the simple (multiplicity 1) synchronization is obtained.

Example 5.2. (Extremal synchronization). Extremal synchronization is understood as a kind of behavior of scalar processes when the processes take their extremum (i.e., maximum or minimum) values simultaneously or with a certain time-shift [75, 111]. The synchronization index in this case is defined as the time of passing through the last extremum: $C_t = t^*(t)$ $(i = 1, 2, \ldots, k)$, where $t^*(t)$ is the time when the process $x(t)$ takes its last extremum value in the interval $0 \leq s \leq t$. In this case the intervals between the first extremum times of $x^i(t)$ and $x^1(t)$ can play the role of the time-shifts τ_i. For vector-valued processes extremal synchronization of any component of the vector $x^{(i)}(t)$ or of any scalar function of $x^{(i)}(t)$ can be considered.

Synchronization of such kind is important for a number of chemical or biological systems.

Example 5.3. (Phase synchronization). Systems of phase synchronization (phase locking) are well known in electronics and telecommunications [266, 272]. However, conventional engineering applications deal with periodic processes with constant or periodically changing frequencies. In the 1990s there were growing activities in the synchronization of chaotic processes. It led to appearance of new definitions of phase and phase synchronization suitable for chaotic processes [375]. It is natural to introduce the phase of a chaotic process considering its behavior between time of crossings of a certain surface (Poincaré section). The synchronization index can then be introduced as the value of the phase φ_t of the process $x(t)$ that belongs to the interval from 0 to 2π and defined as follows:

$$C_t[x] = \varphi_t = 2\pi \frac{t - t_n}{t_{n+1} - t_n} + 2\pi n, \, t_n \leq t < t_{n+1}, \quad (5.1.8)$$

where t_n is the time of the nth crossing of the trajectory with the Poincaré surface [375].

For $k = 2$ the choice of comparison function $F_1(\varphi_t) = F_2(\varphi_t) = \varphi_t$, yields the in phase synchronization. Otherwise, one may choose $F_1(\varphi_t) = \varphi_t, F_2(\varphi_t) = \varphi_t + \pi$, that corresponds to the antiphase synchronization.

A slightly more general concept of synchronization may be obtained if the synchronization index is chosen as follows: $C_t = t_*(t)$, where $t_*(t)$ is the latest time not exceeding t, of crossing the Poincaré section see [75]. Such a concept encompasses the cases that differ from phase synchronization because no physically reasonable notion of *phase* can be introduced owing to significant irregularity of the processes. For example, let the Poincaré surface be chosen so that its equation defines zero value of time derivative of a certain scalar function of the process state. Then the corresponding synchronization concept is just the extremal synchronization (see above). The notion of the *phase* can be extended further, see [330].

Example 5.4. (Coordinate synchronization). In the middle of the 1980s the definition of synchronization for aperiodic processes as coincidence of the coordinates of the interacting subsystems was introduced [7, 170]. This definition has become especially popular after the publishing of the paper by L. Pecora and T. Carroll concerning master–slave synchronization of chaotic systems [343]. Obviously, the coordinate synchronization fits the above definition with the following synchronization index $C_t(x_i) = x_i(t)$, where $x_i(t)$ stands for the value of the state vector of the ith subsystem at the time instant t. Comparison functions can be chosen identical: $F_i(x) = x$, $i = 1, \ldots, k$. In the case when the two processes are generated by two systems, the coordinate synchronization means that the point $(x_1(t), x_2(t))$ in the extended state space of two systems belongs to the diagonal set $\{(x_1(t), x_2(t)) : x_1(t) = x_2(t)\}$.

Example 5.5. (Generalized (partial) coordinate synchronization). The coordinate synchronization from the previous example is often called *full* or *identical* to underline coincidence of all phase coordinates of the systems. In practice a more general case may take place, when only a part of all phase coordinates or certain functions of them $G_i(x_i)$ coincide. The corresponding definition was introduced in [384] and termed *generalized synchronization*. Obviously, the generalized synchronization fits the above scheme under the choice $C_t(x_i) = x_i(t)$, and $F_i(x) = G_i(x)$, $i = 1, \ldots, k$. Still more general property is called *cluster synchronization*. It means that there are several groups (clusters) of coinciding variables. The variables within each group oscillate synchronously while synchronism between different groups may be absent [63–65, 352].

Example 5.6. (Lag synchronization). According to the Definition 5.1 time shifts τ_i may be arbitrary. If the attention is focused on their value, more specific types of synchronization may occur. For example consider coordinate synchronization of two systems $(C_t(x_i) = x_i(t),)$ and assume that the time shifts τ_1 and τ_2 are known. Such a synchronization type, meaning a coincidence of shifted in time states of two systems for all t, i.e., fulfillment of the identity $x_1(t) = x_2(t - \tau_1 + \tau_2)$ was called *lag synchronization* with the lag $\tau_1 - \tau_2$ in [376]. Lags (delays) in the controlled systems or controller may create different effects. For example, an approach named *anticipated synchronization* [438] has attracted a lot of attention because of its potential applications for predicting the dynamics of chaotic systems. Anticipated synchronization describes the situation when the slave system becomes synchronized with the future output of the master system by appropriately including some delay times in feedback terms. It is worth noting that such a possibility is known in the control theory since 1957 as *Smith predictor* [334, 413]. For periodic lag-synchronized oscillations of two generators the point $(x_1(t), x_2(t))$ belongs to an ellipse rather than the diagonal in the extended state space and for multiple synchronization case draws *Lissajous curves*. Delayed feedback may be an efficient algorithm for control of chaos [362] as shown in the Section 6.4.3.

5.1.4 Synchronization of systems

Further specification of the above definition is based on specifying the dynamical models of the processes involved and introducing mechanisms of control. Below we will describe this formalized definition and also introduce its "controlled" version, following [75]. In order to formalize the model of the process

5.1 Definitions of synchronization

any definition of a dynamical system can be employed. Following [75] we will use the input-state-output definition which is standard in control theory [222]. Consider k dynamical systems

$$S_i = \{T, U_i, X_i, Y_i, \phi_i, h_i\}, \qquad i = 1, \ldots, k$$

where T is common set of time instances, U_i, X_i, Y_i are sets of inputs, states and outputs, respectively; $\phi_i : T \times X_i \times U_i \to X_i$ are transition maps, $h_i : T \times X_i \times U_i \to Y_i$ are output maps.

First assume that all U_i are just singletons, i.e., inputs may be dropped from formulations.

Suppose l functionals $g_j : \mathcal{Y}_1 \times \mathcal{Y}_2 \times \ldots \times \mathcal{Y}_k \times T \to \mathbb{R}^1, j = 1, \ldots, l$, are given. Here $\mathcal{Y}_i$ are the sets of all functions from T into Y_i, i.e., $\mathcal{Y}_i = \{y : T \to Y_i\}$.

In the sequel, we take as time set T either $T = \mathbb{R}^1 = t : 0 \leq t < \infty$ (continuous time) or $T = 0, 1, 2, \ldots$ (discrete time). For any $\tau \in T$ the *shift operator* $\sigma_\tau : \mathcal{Y}_i \to \mathcal{Y}_i$ is defined as follows $(\sigma_\tau y)(t) = y(t + \tau)$ for all $y \in \mathcal{Y}_i$ and all $t \in T$. Let $x^{(1)}(t), \ldots, x^{(k)}(t)$ be solutions of the systems $S_1, \ldots, S_k$, $x^{(i)}(\cdot) \in X_i$ with initial states $x^{(1)}(0), \ldots, x^{(k)}(0)$, respectively well defined for all $t \in T$.

Definition 5.2. *We call the processes $x^{(1)}(t), \ldots, x^{(k)}(t)$ synchronized with respect to the functionals $g_1, \ldots, g_l$ if*

$$g_j(\sigma_{\tau_1} y_1(\cdot), \ldots, \sigma_{\tau_k} y_k(\cdot), t) \equiv 0, \quad j = 1, \ldots, l \qquad (5.1.9)$$

is valid for all $t \in T$ and some $\tau_1, \ldots, \tau_k \in T$, where $y_i(\cdot)$ denotes the output function of the system S_i: $y_i(t) = h(x_i(t), t), t \in T, i = 1, \ldots, k$.

We say that the processes $x^{(1)}(t), \ldots, x^{(k)}(t)$ are approximately synchronized with respect to the functionals $g_1, \ldots, g_l$, if there are an $\varepsilon > 0$ and $\tau_1, \ldots, \tau_k \in T$ such that

$$|g_j(\sigma_{\tau_1} y_1(\cdot), \ldots, \sigma_{\tau_k} y_k(\cdot), t)| \leq \varepsilon, j = 1, \ldots, l \qquad (5.1.10)$$

for all $t \in T$.

We call the processes $x^{(1)}(t), \ldots, x^{(k)}(t)$ asymptotically synchronized with respect to the functionals $g_1, \ldots, g_l$, if for some $\tau_1, \ldots, \tau_k \in T$

$$\lim_{t \to \infty} g_j(g_j(\sigma_{\tau_1} y_1(\cdot), \ldots, \sigma_{\tau_k} y_k(\cdot), t)) = 0, \quad j = 1, \ldots, l. \qquad (5.1.11)$$

We call the processes $x^{(1)}(t), \ldots, x^{(k)}(t)$ asymptotically approximately synchronized with respect to the functionals $g_1, \ldots, g_l$ if

$$\overline{\lim}_{t \to \infty} |g_j(\sigma_{\tau_1} y_1(\cdot), \ldots, \sigma_{\tau_k} y_k(\cdot), t)| \leq \varepsilon, \quad j = 1, \ldots, l \qquad (5.1.12)$$

for some $\varepsilon > 0$ and for some $\tau_1, \ldots, \tau_k \in T$.

Finally, if the accuracy threshold ε and time shifts $\tau_1, \ldots, \tau_k \in T$ are specified in advance, we speak about synchronization with accuracy ε and time shifts $\tau_1, \ldots, \tau_k \in T$.

In many practical synchronization problems the spaces $\mathcal{Y}_i$ are identical and the functionals $\{g_j\}$ have the form:

$$g_j(y_s(\cdot), y_r(\cdot)) = \text{dist}(J_j(y_s(\cdot)), J_j(y_r(\cdot))),$$

where $r, s = 1, \ldots, k$, $j = 1, \ldots, l$ and $J_j : \mathcal{Y}_i \to \mathcal{J}_j$, is some mapping (synchronization characteristics) which depends on the (output) trajectory $y_i(\cdot)$ of each of the system $S_1, \ldots, S_k$, and $\mathcal{J}_j$ is some metric space. In this case we will talk about *synchronization with respect to the functionals* $\{g_i\}$ *and characteristics* $\{J_j\}$. For example, J_j may be chosen as the jth component of the state of the system, or as the output function evaluated at some specified time instant. Another choice $J_j = <\dot{\varphi}>$ (the average of rotational velocity (frequency) of some coordinate φ) corresponds to the above-mentioned frequency (Huygens) synchronization.

The specific choice of the synchronization characteristics depends on the essence of the mathematical, physical or engineering problem. The same is valid for the phase shifts τ_i and accuracy threshold ε which may be fixed in some problems and may be arbitrary in others.

Remark 5.2. Note that instead of the set of the functionals it is always possible to take one functional which expresses the same synchronization phenomenon. For example, one can take the functional G as follows

$$G(y_1(\cdot), \ldots, y_k(\cdot), t) = \sum_{j=1}^{l} g_j^2(y_1(\cdot), \ldots, y_k(\cdot), t). \tag{5.1.13}$$

In many practical cases the sets U_i, X_i, Y_i are finite-dimensional vector spaces and the systems S_i can be described by ordinary differential equations. First consider the simplest case of disconnected systems without inputs:

$$S_i : \quad \dot{x}_i = F_i(x_i, t), \tag{5.1.14}$$

where $F_i, i = 1, \ldots, k$ are vector fields. Sometimes synchronization may occur in disconnected systems (5.1.14). For example, all precise clocks are synchronized in the frequency sense. More realistic situations arise when several identical systems are subjected to a single excitation (control) action $f(t)$, e.g., cases of vibrational synchronization [73] and noise-induced synchronization [33, 182, 253, 462, 463] described by mathematical model

$$S_i : \quad \dot{x}_i = F(x_i) + f(t). \tag{5.1.15}$$

The model (5.1.15) can be interpreted as the equation of a single system with a set of initial conditions. Then existence of a single synchronous mode corresponds to the convergence property, see below, Section 5.2.1.

The most interesting and important case, however, is synchronization of interconnected systems. In this case the local subsystem models are augmented with interconnection models:

5.1 Definitions of synchronization

$$\begin{cases} \dot{x}_i = F_i(x_i, t) + \tilde{F}_i(x_0, x_1, \ldots, x_k, t), & i = 1, \ldots, k \\ \dot{x}_0 = F_0(x_0, x_1, \ldots, x_k, t) \end{cases} \quad (5.1.16)$$

where the vector field F_0 describes the dynamics of the interconnection system, $\tilde{F}_i$ are vector fields of the interconnections.

A remarkable and widely used observation is that the synchronization may exist, i.e., identity (5.1.9) may be valid in the interconnected system (5.1.16) without any external action, i.e., without inputs. In this case the system (5.1.16) is called *self-synchronized with respect to the functionals* $g_1, \ldots, g_l$. Similar definitions are introduced for approximate and asymptotic self-synchronization. Usually in this case the systems $S_1, \ldots, S_k$ are autonomous.

In many important application cases the interconnections between the systems $S_1, \ldots, S_k$ are weak, for instance, when (5.1.16) can be represented as

$$\begin{cases} \dot{x}_i = F_i(x_i, t) + \mu \tilde{F}_i(x_0, x_1, \ldots, x_k, t), & i = 1, \ldots, k \\ \dot{x}_0 = F_0(x_0, x_1, \ldots, x_k, t) \end{cases} \quad (5.1.17)$$

where μ is a small parameter. Therefore, finding conditions for self-synchronization in systems with small interactions is of special interest. Such conditions were found for a large class of dynamical systems (5.1.17) with time-periodic functions F_i in the right-hand sides [71, 73]. However, in many cases self-synchronization is not observed and the question arises: is it possible to affect, and thus to control the systems in such a way that the goal (5.1.10) or (5.1.11) can be achieved?

The above definitions do not yet include the possibility of controlling the system. Assume for simplicity that all S_i, $i = 0, \ldots, k$ are smooth finite dimensional systems, described by differential equations with a finite-dimensional input:

$$\begin{cases} \dot{x}_i = F_i(x_i, t) + \tilde{F}_i(x_0, x_1, \ldots, x_k, u, t), & i = 1, \ldots, k \\ \dot{x}_0 = F_0(x_0, x_1, \ldots, x_k, u, t) \end{cases} \quad (5.1.18)$$

where $u = u(t) \in \mathbb{R}^m$ is the input (control variable) which has physical meaning.

The problem of *controlled synchronization with respect to the functionals* $g_j, j = 1, \ldots, l$ is to find a control u as a feedback function of the states $x_0, x_1, \ldots, x_n$ and time providing that (5.1.9) holds for the closed-loop system. The problems of *controlled asymptotic synchronization* and *controlled approximate synchronization with respect to the functionals* $g_j, j = 1, \ldots, l$ are formulated similarly, replacing (5.1.9) by (5.1.10) or (5.1.11). Note that that solving controlled synchronization problems makes sense only if synchronization is not observed at the beginning of the controlled system evolution. It means that typical synchronization goals are asymptotic synchronization (5.1.10) or asymptotic approximate synchronization (5.1.12).

Sometimes the goal can be ensured without measuring any variables of the systems, for instance, by time-periodic forcing (vibrational synchronization). In this case control function u does not depend on system states and the problem of finding such a control is called an *open-loop-controlled (asymptotic) synchronization problem*.

However, the most powerful approach assumes the possibility of measuring the states or some function of the system variables. Finding a control function in this case is called a *closed-loop* or *feedback synchronization problem*.

The simplest form of feedback is *static feedback* where the controller equation is as follows

$$u(t) = \mathcal{U}(x_0, x_1, \ldots, x_k, t) \tag{5.1.19}$$

for some function $\mathcal{U} : \mathbb{R}^{n_0} \times \mathbb{R}^{n_1} , \ldots, \mathbb{R}^{n_k} \times \mathbb{R} \to \mathbb{R}^m$

A more general form is *dynamic state feedback*:

$$\dot{w} = W(x_0, x_1, \ldots, x_k, w, t) \tag{5.1.20}$$
$$u(t) = \mathcal{U}(x_0, x_1, \ldots, x_k, w, t) \tag{5.1.21}$$

with $w \in \mathbb{R}^\nu$, $W : \mathbb{R}^{n_0} \times \mathbb{R}^{n_1} \times \ldots \times \mathbb{R}^{n_k} \times \mathbb{R}^\nu \times \mathbb{R} \to \mathbb{R}^\nu$, $\mathcal{U} : \mathbb{R}^{n_0} \times \mathbb{R}^{n_1} \times \ldots \times \mathbb{R}^{n_k} \times \mathbb{R}^\nu \times \mathbb{R} \to \mathbb{R}^m$.

Now the problem of *state feedback synchronization* can be posed as follows. Find a control law (5.1.19), (or (5.1.20), (5.1.21)) ensuring the asymptotic synchronization (5.1.11) in the closed-loop system (5.1.18), (5.1.19) (or respectively, (5.1.18), (5.1.20), (5.1.21)).

In a variety of practical problems complete information about the states of the systems $S_0, S_1, \ldots, S_k$ is not available and only some *output variables*

$$y_s, \quad s = 1, \ldots, r,$$

are available for using in the control law. In case when the S_i are smooth finite-dimensional systems the problem of output feedback synchronization can be posed as follows: find static control law

$$u(t) = \mathcal{U}(y_1, \ldots, y_r, t) \tag{5.1.22}$$

or dynamical control law

$$\dot{w} = W(y_1, \ldots, y_r, w, t) \tag{5.1.23}$$
$$u(t) = \mathcal{U}(y_1, \ldots, y_r, w, t) \tag{5.1.24}$$

where $w \in \mathbb{R}^\nu$, $y_s \in \mathbb{R}^{p_s}$ $W : \mathbb{R}^{p_1} \times \ldots \times \mathbb{R}^{p_s} \times \mathbb{R}^\nu \times \mathbb{R} \to \mathbb{R}^\nu$, $\mathcal{U} : \mathbb{R}^{p_1} \times \ldots \times \mathbb{R}^{p_s} \times \mathbb{R}^\nu \times \mathbb{R} \to \mathbb{R}^m$, such that the goal (5.1.11) in the system (5.1.18), (5.1.22) (or (5.1.18), (5.1.23), (5.1.24)) is achieved.

5.1.5 Discussion

The proposed definitions allow the researcher to formalize different properties of processes and systems intuitively related to synchronization by means of

5.1 Definitions of synchronization

proper choice of the synchronization index and comparison functions. For example, in order to define coordinate synchronization of processes that change in a correlated manner but with different amplitudes one may use the normalized synchronization index:

$$C_t[x] = \frac{x(t)}{\max_{0 \le s \le t} |x(s)|}.$$

If one of the two T-periodic processes is corrupted by an irregular noise, in order to reveal possible synchronization one may use a moving average-based index:

$$C_t[x] = \frac{2}{T} \int_{t-T/2}^{t} x(s) ds.$$

A different way of using moving average allows the elimination of trends. For example, introducing the synchronization index of the form

$$C_t[x] = x(t) - \frac{1}{T} \int_{t-T}^{t} x(s) ds$$

allows the description of a synchronous behavior modulo a linear trend.

The above general definitions can be further generalized. For example, the time-shifts $\tau_i, i = 1, \ldots, k$ are not constant in many practical problems. It seems reasonable to extend the definition of synchronization in order to capture the problems where time-shifts are not constant but tend to some constant values instead (so-called "asymptotic phases"). In such cases one may replace time-shift operators for each process by time change operators defined as follows:

$$(\sigma_{\tau_i}) x(t) = x(t'_i(t)),$$

where $t'_i : T \to T$, $i = 1, \ldots, k$ are some homeomorphisms such that

$$\lim_{t \to \infty} (t'_i(t) - t) = \tau_i. \quad (5.1.25)$$

Note that in [7] a milder condition $\lim_{t \to \infty} (t'_i(t)/t) = 1$, was proposed instead of (5.1.25). Since the definition in [7] admits infinitely large time shifts, it may describe processes that intuitively do not seem synchronized.

The definition of the synchronization may be modified to include the cases when the processes x_i belong to different functional spaces $\mathcal{X}_i$ (e.g., $x_1(t) \in \mathbb{R}^3$ is described by the Lorenz model while $x_2(t) \in \mathbb{R}^2$ obeys the Van der Pol equation). To this end we introduce so-called *precomparison functions* $F'_i : \mathcal{X}_i \to \mathcal{X}$, which map all the processes into a single space. Based on such a transformation the equalities defining the synchronous mode take the following form:

$$F_1\left(C_{t+\tau_1}[F'(x_1)]\right) = F_2\left(C_{t+\tau_2}[F'(x_2)]\right) = \ldots = F_k\left(C_{t+\tau_k}[F'(x_k)]\right) \quad \forall t \geq 0. \tag{5.1.26}$$

Note that for the case of two processes ($k = 2$) the precomparison functions generate a "rectifying" transformation $\mathcal{F}$, introduced in the paper [82] as follows $\mathcal{F} = (F'_1, F'_2)$. Let the synchronization index be chosen as in Example 5.5: $C_t(x_i) = x_i(t)$. Let one of the postcomparison functions be chosen identity. Then the definition of synchronization will coincide with the one given in [82]. Moreover, in this case the second comparison function plays the role of synchronization function of the paper [82]. In summary, the definition of [82] describes a more general concept than the generalized synchronization of Example 5.5, yet still less general than the definition introduced in the Section 5.1.1.

It is worth to note in the conclusion that the introduced definitions not only provides the terminology and the conceptual tools for discussion of different synchronization-like properties, but also allows the question of the applicability area for the term "synchronization" to be addressed. For example, according to the above definition the property defined as reduction of the fractal dimension of the overall process with respect to the sum of the dimensions of its components [255] is not a synchronization property, because it is defined via characteristics (dimensions) that do not depend on the behavior of the processes in time. In this case the terms "ordering" or "synergy" would seem more preferable.

Similarly, it seems not appropriate to term two processes as synchronized if they are strongly correlated: $|\varrho(x_1, x_2)| > 1 - \varepsilon$, where

$$\varrho(x_1, x_2) = \frac{\langle x_1 \cdot x_2 \rangle}{\sqrt{\langle x_1^2 \rangle \cdot \langle x_2^2 \rangle}}$$

is the correlation coefficient.

Finally, introduction of the scalar synchronization measure (e.g., (5.1.6)) opens the possibility for systematic design of controlled synchronization systems where the synchronization mode is created or modified artificially. Such design may be based, for example, upon the speed-gradient algorithms

$$u(t) = -\gamma \nabla_u \dot{Q}_t, \quad \gamma > 0,$$

where $u(t)$ is the vector of the controlling parameters or variables, see [26, 76, 164] and Section 2.4.2. The issue of controlled synchronization systems design will be addressed in the next section.

5.2 Controlled synchronization design

The existing methods of dynamical systems theory and control theory can be used to analyze and design a variety of synchronization systems. Below we consider some design problems of asymptotic coordinate synchronization understood as fulfillment of the relation

$$\lim_{t\to\infty} |x_1(t) - x_2(t)| = 0, \qquad (5.2.27)$$

where x_1, x_2 are state vectors of the subsystems to be synchronized.

Control methods for other types of synchronization can be found in the literature. For example, classical phase-locked loops [272] can be interpreted as phase synchronization control systems and studied by control theory tools [266]. Control method for phase synchronization in coupled complex oscillators was proposed in [66]. Multiple synchronization control was studied in [76]. Synchronization control methods for spatiotemporal systems can be found, e.g., in [79, 238, 450].

5.2.1 Synchronization and convergence

In the stability theory of dynamical system an analog of the synchronization property known as *convergence* is well known.

A system of differential equations is called *convergent* [117], if it has a unique bounded solution which is globally asymptotically stable, i.e., all motions of a system converge to a certain limit mode. Consider two identical systems:

$$\dot{x}_1 = F(x_1, t), \quad \dot{x}_2 = F(x_2, t). \qquad (5.2.28)$$

If $x_1(t)$ and $x_2(t)$ are arbitrary solutions of the systems (5.2.28), then they can be interpreted as two solutions of the single system $\dot{x} = F(x, t)$ at different initial conditions. Therefore the convergence of the system $\dot{x} = F(x, t)$ implies coordinate synchronization of identical systems (5.2.28). The sufficient condition for convergence (and for synchronization) is the so-called Demidovich condition [117]: all eigenvalues of the symmetrized Jacobi matrix of the system $J(x,t) = \frac{\partial F(x,t)}{\partial x} + \left[\frac{\partial F(x,t)}{\partial x}\right]^T$ should be uniformly negative:

$$\lambda_i\left(\frac{\partial F(x,t)}{\partial x} + \left[\frac{\partial F(x,t)}{\partial x}\right]^T\right) \leq -\delta \qquad (5.2.29)$$

for some $\delta > 0$ and for all $t \geq\geq 0$, $x \in \Omega$, where Ω is a set covering the trajectory $x(t)$. The condition (5.2.29) can be understood as a strengthened stability of free motions of the system. Oscillatory systems, however, are often close to the stability border. Moreover, they may travel beyond the stability region, as in the case of chaotic oscillations. In such cases the problem of controlled synchronization arises and synchronization system design is needed.

One way to achieve synchronization is to achieve fulfillment of the Demidovich condition by applying an external action. However, it was observed in some cases that it is not necessary to ensure stability in the whole state space in order to achieve synchronization. Synchronization may occur even the system trajectories spend sufficiently large part of time in the stability region. The related problem of *capture* (called also entrainment or external synchronization) of trajectories of nonlinear oscillatory system driven by an external harmonic signal is a classical problem of oscillation theory [30, 34, 73, 253, 433]. Recently it was discovered that noise or chaotic signals can also lead to synchronized behavior of decoupled systems [33, 123, 182, 249, 250, 253, 462, 463].

Perhaps, the first mathematical treatment of convergence (capture) phenomenon for unstable systems was made by G.A. Leonov [264]. In [264] a class of Lurie systems dissipative yet unstable near origin was considered and the frequency-domain conditions ensuring that for large excitation amplitude the system trajectory spends enough time in the stability region to ensure synchronization were proposed. The approach was extended to other classes of systems and nonperiodic excitations [108, 156, 174]. Properties of more weak synchronization property – phase synchronization – were investigated in [349, 462, 463].

5.2.2 Synchronization and stabilization

Let us illustrate some of the peculiarities of the controlled synchronization for the special case when the models of the synchronized systems are linear both in states and in controls:

$$\dot{x}_1 = Ax_1 + f(t) + Bu_1, \qquad (5.2.30)$$

$$\dot{x}_2 = Ax_2 + f(t) + Bu_2, \qquad (5.2.31)$$

where u_1, u_2 are controlling actions, $f(t)$ is the external action or disturbance which is assumed to be a bounded function of time. The matrices of coefficients A, B are of size $n \times n$, $n \times m$, respectively. In this case the synchronization error $e = x_1 - x_2$ obeys the linear equation:

$$\dot{e} = Ae + B(u_1 - u_2) \qquad (5.2.32)$$

and the problem of asymptotic synchronization of the subsystems (5.2.30) and (5.2.31) is reduced to asymptotic stabilization (establishing asymptotic stability for the error equation (5.2.32). Obviously, the dynamics of the synchronization error depends only on the difference between controls $u = u_1 - u_2$. Therefore, it is sufficient to study the case of only one control and to seek a linear feedback control of the form $u_1 = Ke, u_2 = 0$. For such a case the control does not influence the system (5.2.31) and its motions play the role of command (reference) motions for the system (5.2.30). The systems with reference models are well known in the control theory. In physical literature they are called *master–slave* or *drive-response* systems. To achieve the synchronization goal the feedback matrix K should be chosen in such a way that the matrix $A + BK$ is to be Hurwitz (stable), i.e., all the roots of its characteristic polynomial $\det(\lambda I - A - BK)$ would have negative real parts. It is a well-known fact of the control theory that if the pair of matrices A, B is controllable[1] then it is possible to choose the feedback matrix K providing any desired eigenvalues of the matrix $A + BK$. In addition, it is easy to show that if the matrix A does not have "right" eigenvalues, then the synchronizing control can be arbitrarily small (Indeed, to achieve synchronization one needs only to shift the eigenvalues lying in the imaginary axis to the left, and that can be achieved by a small control owing to continuous dependence of the coefficients of a polynomial from its roots).

Similar conclusions hold in the symmetric case $u_1 = -u_2 = u/2$, when control acts on both subsystems. Such problems are not typical for control theory, since in this case the desired motion of the overall system is not predefined. In the control theory a related problem was studied: the so-called *coordinated control* [299]. In the coordinated control problem two controls (u_1, u_2) are used to achieve an additional goal: stabilization of the base (averaged) motion $x_1 + x_2)/2$. It is interesting that the same methods can be used both for coordinated control and for synchronization, since the error equations coincide.

Thus, the problem of feedback design for synchronization of linear systems with full state measurements is well understood. However, the problems with incomplete measurements are much more complicated. Let, instead of the state vectors x_1, x_2, only the output vectors $y_1 = Cx_1$, $y_2 = Cx_2$, where C is rectangular $l \times n$-matrix, $l < n$, be available for measurement. The existing criteria for output feedback stabilization are rather involved. No simple final solution of the problem is obtained even for linear systems. Below we formulate some simple sufficient conditions. To this end let us introduce the transfer matrix of a linear system $\dot{x} = Ax + Bu$ in a standard manner: $W(\lambda) = C(\lambda I - A)^{-1}B$. Assume for simplicity that $l = m = 1$, i.e., both input and output are scalars. In this case $W(\lambda) = b(\lambda)/a(\lambda)$, where $b(\lambda)$, $a(\lambda)$ are polynomials of degree n_1, n, respectively, $n_1 < n$, $a(\lambda)$ is the characteristic polynomial of the matrix A. The error equation can be transformed to the differential equation of nth order $a(p)y = b(p)u$, where $p = d/dt$ is symbol of differentiation. Without further loss of generality we may assume that the coefficient a_n at λ_n is equal to one, $a_n = 1$.

A sufficient condition for stabilizability of a linear system by output feedback $u = Ky$ was established as early as in the end of the 1940s by M.V. Meerov (see [293] and references therein): there exists a stabilizing feedback K if the polynomial $b(\lambda)$ is Hurwitz (all its roots have negative real parts), and the value $d = n - n_1$ (difference between the degrees of denominator and numerator of the transfer function, the so-called *relative degree of the system*) is equal to one or two, for $d = 2$ the coefficient at λ^{n-1} being positive $a_{n-1} > 0$. If the Meerov condition holds, then any K with sufficiently large

---

[1] The necessary and sufficient condition for controllability of A, B is the so-called Kalman criterion $\text{rank}\{B, AB, \ldots, A^{n-1}B\} = n$.

absolute value and the sign opposite to the sign[2] of the coefficients of the numerator $b(\lambda)$ of the transfer function $W(\lambda)$: $b_0 K < 0$. Besides, it was established by Meerov that if $d \geq 3$ or $d = 2$, but $a_{n-1} \leq 0$, then stabilization of an unstable system by an output feedback is impossible. In such a case the output feedback synchronization problem also cannot be solved.

5.2.3 Synchronization and observers

More possibilities for synchronization arise when the structural restrictions for control are absent in one of subsystems. This is the case when one of the subsystems to be synchronized should be implemented as an analogue or digital device. Let the whole system be described by the equations

$$\dot{x}_1 = Ax_1 + f(t), \quad y_1 = Cx_1, \tag{5.2.33}$$

$$\dot{x}_2 = Ax_2 + u(t). \tag{5.2.34}$$

The problem of controlled coordinate synchronization in this case is to find a controlling function $u = \mathcal{U}(y_1, x_2, t)$, ensuring the relation (5.2.27) in the closed-loop system. It can be interpreted as reconstruction of the state x_1 of the system (5.2.33) by means of its estimate x_2. Such a problem is well studied in the field of control theory where it is called the *observation problem*. Its solution is provided with the so-called *linear observer*

$$\dot{x}_2 = Ax_2 + K(y_1 - Cx_2) + f(t), \tag{5.2.35}$$

where matrix K is to be found. Obviously, the observation error $e(t) = x_1(t) - x_2(t)$ obeys the equation

$$\dot{e} = (A - KC)e.$$

It is well known that the eigenvalues of the matrix $A - KC$ can be chosen arbitrarily by means of the proper choice of the matrix K if the so-called *observability condition* holds.[3]

The synchronization systems based on the linear error equation described above possess such important properties as roughness and robustness. *Roughness* means that the small changes in the system model (taking into account nonidentity or nonlinearity of subsystems, interactions, external actions, etc.) result in only small changes in the system behavior. In the context of stability it coincides with *structural stability* property. *Robustness* is the stronger property requiring additional quantitative estimates for changes in the system behavior as a function of changes of the system model. For example, it can be shown that if the functions describing these additional changes of the model appear additively in the right-hand sides of the system equations and their

---

[2] All the coefficients of Hurwitz polynomial $b(\lambda)$ are of the same sign.
[3] Observability condition for the pair of matrices A, C is $\text{rank}\{C, A^\mathrm{T} C, \ldots, (A^\mathrm{T})^{n-1} C\} = n$. It is dual to controllability condition

norms are bounded by some quantity Δ, then the limit (for $t \to \infty$) norm of the error is bounded by the value $R\Delta$ for some $R > 0$, where R does not depend on Δ. It means that the error is of the same order as the disturbing factors.

A peculiarity of the roughness and robustness properties in this problem is in that the roughness and robustness take place with respect to synchronization error rather than with respect to the behavior of each subsystem. The perturbed subsystems may loose stability or some of their variables may grow infinitely (e.g., pendulums may turn into a rotation mode) but deflection from the synchronous mode (synchronization error) will remain bounded.

5.2.4 Synchronization of affine nonlinear systems

In physics the most interesting synchronization problems arise in nonlinear systems. Let us briefly describe a few methods of synchronization design in nonlinear systems. For the sake of simplicity consider two systems described by the models affine in control.

$$\begin{aligned} \dot{x}_1 &= f_1(x_1) + g_1(x_1)u_1, \\ \dot{x}_2 &= f_2(x_2) + g_2(x_2)u_2. \end{aligned} \quad (5.2.36)$$

Initially the systems are separate, i.e., not linked. Let us pose the problem of coordinate synchronization design as follows. Find a control algorithm

$$u_i = U_i(x_1, x_2), \quad i = 1, 2, \quad (5.2.37)$$

such that the control goal (5.2.27) is attained. The solution of the problem is trivial in the case when the right-hand sides of (5.2.36) can be changed independently and arbitrarily, i.e., if $m = n$, $g_1(x_1) = g_2(x_2) = I_n$, where I_n is identity $n \times n$-matrix. Then, taking, for instance, $u_1 = 0, u_2 = K(x_1 - x_2)$, where $K > 0$ is the gain coefficient, we obtain the error equation in the form

$$\dot{e} = f(x_1(t)) - f(x_1(t) - e) - Ke, \quad (5.2.38)$$

where $x_1(t)$ is a given function of time, satisfying the first equation in (5.2.36) for $u_1 = 0$. Suppose that the Jacobian matrix $A(x) = \dfrac{\partial f}{\partial x}(x)$ is bounded in some region Ω, containing the initial condition $x(0)$ of (5.2.36). Then the eigenvalues of the symmetric matrix $A(x) + A^T(x) - 2KI_n$ have negative real parts for $x \in \Omega$ for sufficiently large $K > 0$. It follows from Demidovich condition that the system possesses convergence property in Ω, , i.e., all its solutions starting in Ω, converge to a unique bounded solution for $t \to \infty$. Since the solution $e(t) \equiv 0$ is nothing but such a solution, all other solutions tend to it when $t \to \infty$. Therefore the synchronization of two subsystems is observed if the coupling strength between them increases. At the same time the behavior of each subsystem may remain complex, e.g., chaotic. Note that smoothness of the right-hand sides and existence of the Jacobian matrix are not needed for synchronization: it is sufficient to verify the Lipschitz condition: $|f(x_1) - f(x_2)| \le L|x_1 - x_2|$ for some $L > 0$.

Similarly, for Lipschitz systems it is possible to design a nonlinear observer, so-called *high-gain observer*:

$$\dot{x}_2 = f_2(x_2) + Kg_2(x_2)(y_1 - Cx_2), \qquad (5.2.39)$$

where $y_1 = Cx_1$. Applicability conditions for high-gain observers can be found, e.g., in [268]. Particularly, it is required that the Lipschitz constant L is sufficiently small.

5.2.5 Pecora–Carroll scheme

An interesting observer design scheme was proposed by L. Pecora and T. Carroll in 1990 and soon has become very popular [343]. The scheme is applicable if the dynamics equations can be split into two groups, corresponding observed variables y_1 and nonobserved variables z_1 in the following way:

$$\dot{y}_1 = F_y(y_1, z_1), \qquad (5.2.40)$$

$$\dot{z}_1 = F_z(y_1, z_1), \qquad (5.2.41)$$

the second subsystem (5.2.41) possessing the convergence property with respect to z_1. Then the vector y_1 may be introduced directly into the observer with the state vector z_2, described by the equation

$$\dot{z}_2 = F_z(y_1, z_2). \qquad (5.2.42)$$

It follows from the convergence property that $z_1(t) - z_2(t) \to 0$ as $t \to \infty$ and, therefore the vector $\mathrm{col}(y_1, z_2)$ may serve as the estimate of the state vector of the system (5.2.40), (5.2.41). As for applicability condition for this scheme, the Demidovich sufficient condition of convergence can be used: the eigenvalues of the matrix

$$\frac{\partial F_z(y, z)}{\partial z} + \left[\frac{\partial F_z(y, z)}{\partial z}\right]^\mathrm{T},$$

are uniformly negative for all y, z. Such a condition is easier to check than negativity of conditional Lyapunov exponents, proposed in [343]. Justification for a more general case can be found in [164].

The Pecora–Carroll scheme can be represented as the limit case of the high-gain observer, having the structure

$$\dot{y}_2 = F_y(y_2, z_2) + K(y_1 - y_2), \qquad (5.2.43)$$

$$\dot{z}_2 = F_z(y_2, z_2). \qquad (5.2.44)$$

Indeed, rewriting (5.2.43) in the form

$$\varepsilon \dot{y}_2 = \varepsilon F_y(y_2, z_2) + (y_1 - y_2), \qquad (5.2.45)$$

where $\varepsilon = 1/K$ is the small parameter, one can notice that the system (5.2.43), (5.2.44) belongs to the class of so-called singularly perturbed systems, exhibiting fast and slow motions. Make a standard reduction to extract a reduced system, describing slow motions. To this end one should set $\varepsilon = 0$ and obtain the reduced system in the form: $y_1 = y_2$, $\dot{z}_2 = F_z(y_1, z_2)$ coinciding with (5.2.42). Since the fast subsystem (5.2.43) is asymptotically stable for sufficiently large $K > 0$, its solution $y_2(t)$ is close to $y_1(t)$ for sufficiently large $K > 0$ and $t > 0$, i.e., the dynamics of the observer (5.2.43), (5.2.44) are determined by the dynamics of the reduced system (5.2.42).

Let us show an example of application of the Pecora–Carroll scheme to synchronization of the Lorenz system.

Example 5.7. The Pecora–Carroll scheme was applied by a number of authors to signal transmission with a chaotic carrier. In the pioneering work of K. Cuomo, A. Oppengeim, and S. Strogatz [112] a signal transmitter was based on the Lorenz system transformed after rescaling to the following equations

$$\begin{cases} \dot{u} = \sigma(v - u), \\ \dot{v} = ru - v - 20uw, \\ \dot{w} = 5uv - bw. \end{cases} \qquad (5.2.46)$$

The choice of parameter values $\sigma = 16$, $r = 45.6$, $b = 4.0$ provides the system with a chaotic behavior.

The equations of the receiver are chosen according to the Pecora–Carroll scheme as follows

$$\begin{cases} \dot{u}_s = \sigma(v_s - u_s), \\ \dot{v}_s = ru - v_s - 20uw_s, \\ \dot{w}_s = 5uv_s - bw_s. \end{cases} \qquad (5.2.47)$$

The equations of (5.2.47) are similar to (5.2.46), with the only exception: the right-hand side of (5.2.47) depends on the variable u, instead of u_s, which is considered as the receiver input signal coming from the transmitter. The system (5.2.46), (5.2.47) fits the Pecora–Carroll scheme for $y = u$, $z = (v, w)$.

Using the Lyapunov function $V = V(e_2, e_3) = 0.5e_2^2 + 4e_3^2$ a simple proof is given in [112] for the fact that the state variables of the systems (5.2.46) and (5.2.47) are synchronized, i.e., the vector of state error tends to zero. To prove this result, the error equation can be written as follows:

$$\begin{cases} \dot{e}_2 = -e_2 - 20ue_3, \\ \dot{e}_3 = 5ue_2 - be_3. \end{cases} \qquad (5.2.48)$$

Calculating the derivative of the function V along trajectories of the system (5.2.48) one arrives at the expression

$$\dot{V} = e_2(-e_2 - 20ue_3) + 4e_3(5ue_2 - be_3) == -e_2^2 - 4be_3^2. \qquad (5.2.49)$$

Therefore, $\dot{V} \leq -kV$, where $k = min\{2, 2b\} > 0$ does not depend on the value of the input signal $u = u(t)$, and the error variables $e_2(t), e_3(t)$ tend to zero exponentially. In view of $\dot{e}_1 = \sigma(e_2 - e_1)$, $\sigma > 0$, the variable e_1 also tends to zero exponentially, i.e., (5.2.47) may serve as asymptotic observer for (5.2.46).

In order to transmit a binary message the coefficient b of the transmitter (5.2.46) was changed to attain the value $b = 4.4$, corresponding to the message value "1", while the initial value $b = 4.0$ corresponded to the binary "0". When the value of b in (5.2.46) is changed to $b = 4.4$, the level of the discrepancy signal $e = u - u_s$, in the system (5.2.47) is increased dramatically. It allows the detection of the transmitted message. □

5.2.6 Synchronization and speed-gradient

To design control algorithms for synchronization problems one may use the speed-gradient algorithms, see section 2.4.2.

Let the controlled system be described by the equations: (5.2.36). Introduce the goal functional

$$Q(x) = \frac{1}{2}|x_1 - x_2|^2 \qquad (5.2.50)$$

and evaluate the rate of its changing along trajectories of the system:

$$\dot{Q}(x) = (x_1 - x_2)^T (f_1 + g_1 u_1 - f_2 - g_2 u_2).$$

Then evaluate the speed gradient

$$\frac{\partial \dot{Q}}{\partial u_1} = (x_1 - x_2)^T g_1,$$
$$\frac{\partial \dot{Q}}{\partial u_2} = -(x_1 - x_2)^T g_2,$$

and write down the speed-gradient algorithm for the functional (5.2.50) as follows:

$$\begin{aligned} u_1 &= -\gamma(x_1 - x_2)^T g_1(x_1), \\ u_2 &= -\gamma(x_2 - x_1)^T g_2(x_2). \end{aligned} \qquad (5.2.51)$$

As it was mentioned in [137] all the above synchronization schemes can be obtained as special cases of the equation (5.2.51). For example, linear feedback synchronization algorithm (5.2.38) can be obtained letting $g_1(x_1) = I_n$, $g_2(x_2) = 0$, $\gamma = K$. The high-gain observer (5.2.39) can be obtained, letting $g_1(x_1) = 0$, $g_2(x_2) = g_2 C$, $\gamma = K$, and the Pecora–Carroll scheme (5.2.42) for the case when the output y_1 appears in (5.2.42) linearly ($F_z(y_1, z_2) = f(z_2) + y_1 \bar{g}(z_2)$) letting $g_1(x_1) = 0$, $g_2(x_2) = gC$, $\gamma = 1$.

The conditions ensuring synchronization for the algorithm (5.2.51) can be derived from general results establishing attainment of the goal in the speed-gradient systems, see [135, 157, 164].

5.3 Adaptive synchronization

5.3.1 Problem formulation

The dynamics of many physical system depend on a number of parameters which are unknown to the system researcher. To design synchronization algorithms in such cases the adaptive control approach can be used. Adaptive control is based on extracting information about unknown parameters during the process of a normal system functioning. It is well-developed branch of the control theory both for linear and for nonlinear systems [132, 135, 157, 245]. However, application of adaptive control to synchronization problems has been systematically studied only recently. Below a general approach to adaptive synchronization based on the so-called *passification method* is presented, following [25, 137, 159].

Consider k-interconnected subsystems whose dynamics are described by equations

$$\dot{x}_i = F_i(x_1, \ldots, x_k, u, \theta, t) \quad i = 1, \ldots, k, \tag{5.3.52}$$

where $\theta \in \mathbb{R}^M$ is the vector of unknown parameters. Let a nonnegative goal function $Q(x_1, \ldots, x_k, t)$ be defined in such a way that its small values correspond to the achievement of the synchronous mode. The choice of the goal function may also reflect the desired type of synchronization. For instance, the goal function can be chosen in the form (5.1.6), (5.1.13), or (5.2.50). The problem is to find the adaptive control algorithm of the form

$$u = U(x_1, \ldots, x_k, t, \theta),$$

where $\theta \in \mathbb{R}^M$ is the vector of adjustable parameters and the adaptation algorithm of the form

$$\dot{\theta} = \Theta(x_1, \ldots, x_k, t, \theta),$$

such that the control goal

$$\lim_{t \to \infty} Q(x_1(t), \ldots, x_k(t), t) = 0 \tag{5.3.53}$$

is achieved for all $\theta \in \Xi$, where Ξ is the set of all possible values of θ.

Note that the right-hand sides F_i in (5.3.52) may be different and, therefore, the posed problem encompasses the case of nonidentical subsystems which attracts the most interest in the controlled synchronization problems.

5.3.2 Adaptive synchronization of two subsystems

Below the case of two subsystems will be examined in detail. Let $k = 2$, $x_i \in \mathbb{R}^n$ and the control action be scalar: $u \in \mathbb{R}^1$. Let the standard coordinate synchronization goal be specified:

$$\lim_{t \to \infty} |x_1(t) - x_2(t)| = 0. \tag{5.3.54}$$

Introduce the error vector $e = x_1 - x_2$ and subtract the equation of the second subsystem from the equation of the first one to obtain the equation for error. Assume that it is possible to separate linear and nonlinear parts in the error equation and represent it in the form:

$$\dot{e} = Ae + B\sum_{i=1}^{N} \theta_i \varphi_i(x_1, x_2, t) + Bu, \qquad (5.3.55)$$

where A is a constant $n \times n$-matrix; B is a constant n-dimensional vector; θ_i are constant but unknown coefficients and functions φ_i are known or available for measurements. It means that the parametrization is assumed to be (A) linear and (B) matched: both unknown parameters and control appear linearly in the error equation and, moreover, they appear proportionally to the constant vector B. A typical class of the systems when the introduced assumptions are satisfied corresponds to the case when both nonlinearities and control appear only in one of the system equations.

The error model (5.3.55) encompasses both conventional situation for control design problems when control appears only in one of the two subsystems (5.3.52), and a nonconventional case when control may influence both subsystems. In the latter case the limit motion of the controlled system is, generally speaking, unknown, even if the error has achieved zero.

Let, in addition to the functions $\varphi_i(x_1, x_2, t)$, the output variables $y_i = Cx_i$, $i = 1, 2$ be available for measurement. The problem is to design control algorithm and adaptation algorithm ensuring convergence of the system error to zero, i.e., ensuring the goal (5.3.54).

To solve the problem the speed-gradient method can be employed. Define the main loop of the system in the form

$$u = -\hat{\theta}_0(y_1 - y_2) + \sum_{i=1}^{N} \hat{\theta}_i \varphi_i(x_1, x_2, t), \qquad (5.3.56)$$

where $\hat{\theta}_0, \hat{\theta}_i$ are some adjustable parameters. Such a choice is motivated by the hope that it is able to solve the problem in principle. Indeed, with such a choice there exist the "ideal" values of adjustable parameters $\hat{\theta}_i$, $i = 1, \ldots, N$, such that the control goal is achieved. Obviously, one can choose the values

$$\hat{\theta}_{i_*} = -\theta_i, \quad i = 1, \ldots, N, \qquad (5.3.57)$$

as "ideal" parameters. Then, substituting the above values of parameters $\hat{\theta}_{i_*}$ and control u into the error equation (5.3.55), we achieve cancellation of all the nonlinearities and the resulting equation takes form

$$\dot{e} = [A - \theta_0 BC]e. \qquad (5.3.58)$$

If there exists θ_{0_*} such that the equation (5.3.58) is asymptotically stable then the feedback control law (5.3.56), (5.3.57) ensures synchronization in principle. However one cannot use the control law (5.3.56), (5.3.57) since it depends on the unknown parameters.

To design the adaptation algorithm let us use the speed-gradient method. Taking into account that the system contains a linear in e part, choose the quadratic function $Q(e) = e^T P e$, $e = x_1 - x_2$, where $P = P^T > 0$ is some symmetric positive-definite matrix. Evaluating the rate of changing the function Q along trajectories of the system (5.3.55), and then its gradient with respect to adjustable parameter, we obtain

$$\dot{Q} = e^T P \dot{e} = e^T P \left[Ae + B \sum_{i=1}^{N} \theta_i \varphi_i + Bu \right] =$$

$$= e^T P \left[A - \hat{\theta}_0 BC \right] e + e^T PB \sum_{i=1}^{N} (\theta_i - \hat{\theta}_i) \varphi_i,$$

$$\frac{\partial \dot{Q}}{\partial \hat{\theta}_0} = -e^T PB(y_1 - y_2), \quad \frac{\partial \dot{Q}}{\partial \hat{\theta}_i} = -e^T PB \varphi_i.$$

To achieve a feasible solution of the problem, one needs to ensure that all the variables in the algorithm are available for measurement. By assumption, the functions $\varphi_i(x_1, x_2, t)$ are available for measurement. Then one needs to guarantee availability of the value $e^T PB$, which is a linear combination of the state error variables. Obviously, this variable can be measured if it is a linear combination of some outputs error variables: $e^T PB = (y_1 - y_2)^T g = e^T C^T g$ for some number g, which is equivalent to the relation $PB = C^T g$. If such a relation holds, then the adaptation algorithm, obtained by the speed-gradient method in differential form takes the following form

$$\dot{\hat{\theta}}_i = -\gamma_i (y_1 - y_2) \varphi_i(x_1, x_2, t), \quad i = 1, \ldots, N, \qquad (5.3.59)$$

$$\dot{\hat{\theta}}_0 = -\gamma_0 (y_1 - y_2)^2, \qquad (5.3.60)$$

where γ_i, $i = 1, \ldots, N$ are adaptation gains. Adaptation gains may have arbitrary absolute value and their sign should coincide with the sign of g.

In a more general case when the measured outputs are l-dimensional vectors, the gain $g \in \mathbb{R}^l$ is also a vector and the adaptation algorithm takes the form

$$\dot{\hat{\theta}}_i = -\gamma_i g^T (y_1 - y_2) \varphi_i(x_1, x_2, t), \quad i = 1, \ldots, N, \qquad (5.3.61)$$

$$\dot{\hat{\theta}}_0 = -\gamma_0 [g^T (y_1 - y_2)](y_1 - y_2), \qquad (5.3.62)$$

where $\gamma_i > 0$, $i = 1, \ldots, N$.

5.3.3 Conditions for control goal achievement

Next we will derive the conditions ensuring achievement of the goal (5.3.54). We need some definitions and results of control theory.

Definition 5.3. ([[135]]) A linear system $\dot{x} = \bar{A}x + \bar{B}u$, $y = \bar{C}x$ with the transfer matrix $W(\lambda) = \bar{C}(\lambda I - \bar{A})^{-1}\bar{B}$, where u, $y \in \mathbb{R}^l$ and $\lambda \in \mathbb{C}$ is called *minimum-phase* if the polynomial $\varphi(\lambda) = \det(\lambda I - \bar{A}) \det W(\lambda)$ is Hurwitz. The system is called *hyper-minimum-phase* if it is *minimum-phase* and the matrix $\bar{C}\bar{B} = \lim_{\lambda \to \infty} \lambda W(\lambda)$ is symmetric and positive definite.

Note that for $l = 1$ the system of nth order is hyper-minimum-phase, if the numerator of its transfer function is the Hurwitz polynomial of degree $n - 1$ with positive coefficients (it is equivalent to the case $d = 1$ in Meerov conditions, see section 5.3.

Lemma 5.1 (Passification lemma) [132, 135, 157]. Let the matrices $\bar{A}, \bar{B}, \bar{C}, g$ of size $n \times n$, $n \times m$, $l \times n$, $m \times l$ be given and the full-rank condition $\operatorname{rank}(\bar{B}) = m$ holds. Then for existence of a positive-definite $n \times n$-matrix $P = P^T > 0$ and a $l \times m$-matrix θ_* such that

$$PA_* + A_*^{\mathrm{T}}P < 0, \quad P\bar{B} = \bar{C}^{\mathrm{T}}g^{\mathrm{T}}, \quad A_* = \bar{A} + \bar{B}\theta_*G\bar{C}$$

it is necessary and sufficient, that the system $\dot{x} = \bar{A}x + \bar{B}u$, $y = G\bar{C}x$ would be hyper-minimum-phase.

It follows from the lemma, see, e.g., [157], that for hyper-minimum-phase system there always exists an output feedback $u = \theta_0 Gy + \bar{u}$, where $\bar{u}$ is a new auxiliary input, such that the closed-loop system is strictly passive with respect to the output $\bar{y} = Gy$. Besides, the storage function can be chosen as a quadratic form: $V(x) = x^{\mathrm{T}}Px$. In other words, the hyper-minimum-phase property is necessary and sufficient for passifiability of a linear system by an output feedback.

Definition 5.4. Vector-function $f : [0, \infty) \to \mathbb{R}^m$ is called *persistently exciting* (PE) *on the interval* $[0, \infty)$, if it is bounded on $[0, \infty)$ and there exist $\alpha > 0, T > 0$ such that

$$\int_t^{t+T} f(s)f(s)^T ds \geq \alpha I_m$$

for all $t \geq 0$. □

The essence of the notion of persistent excitation is in that a persistently exciting vector-function does not approach any hyperplane in the m-dimensional space.

Lemma 5.2 (Persistent excitation lemma) [132, 157]. Consider vector-functions $f, \tilde{\theta} : [0, \infty) \to \mathbb{R}^m$. Let $\tilde{\theta}(t)$ be a continuously differentiable function, such that $d\tilde{\theta}(t)/dt \to 0$ for $t \to \infty$. Let f be persistently exciting and $\tilde{\theta}(t)^{\mathrm{T}} f(t) \to 0$ for $t \to \infty$. Then $\tilde{\theta}(t) \to 0$.

Now it is possible to formulate conditions providing adaptive synchronization. A concise formulation has the form of a theorem.

Theorem 5.1. [159]. *Let the trajectories of the systems to be synchronized (5.3.52) for* $k = 2$, *affected by control (5.3.56) be bounded and the linear system with the transfer function* $W(\lambda) = gC(\lambda I - A)^{-1}B$ *be hyper-minimum-phase.*

Then all the trajectories of the system (5.3.55), (5.3.56),(5.3.59), (5.3.60) are bounded and the synchronization goal (5.3.54) holds. In addition, if the PE condition holds for the vector-function $\mathrm{col}\,(\varphi_1(x_1, x_2, t), \ldots, \varphi_N(x_1, x_2, t))$, *then the adjustable parameters tend to their ideal values:*

$$\lim_{t \to \infty} \left(\hat{\theta}(t) - \theta \right) = 0. \tag{5.3.63}$$

The main condition of the theorem is the hyper-minimum-phaseness of the error equation which ensures the possibility of its passification by an output feedback. By Lemma 5.1, the hyper-minimum-phaseness means that the real parts of the transfer function $W(\lambda) = gC(\lambda I - A)^{-1}B$ numerator zeros are negative. In addition, it needs the *relative degree one condition*: the relative degree[4] of the linear part of the transmitter should be equal to one. The hyper-minimum-phaseness is equivalent to asymptotic stability of zero dynamics [135, 157] which, in turn, is analogous to negativity of transverse Lyapunov exponents condition introduced in [343]. More information about passivity, passification, and hyper-minimum-phaseness of systems can be found in [157, 164].

Proof of Theorem 5.1. Consider the Lyapunov function candidate of the form

$$V(x, \hat{\theta}_0, \hat{\theta}) = \frac{1}{2} e^T P e + \sum_{i=0}^{N} \frac{1}{2\gamma_i} |\hat{\theta}_i - \theta_i|^2 + |\hat{\theta}_0 - \theta_{*0}|^2/(2\gamma_0), \tag{5.3.64}$$

where a matrix $P = P^T > 0$ and a number θ_{*0} are to determine. Calculation of the value of $\dot{V}$ shows that the inequality $\dot{V} < 0$ holds for $e \neq 0$ if and only if the following relations hold:

$$\begin{cases} \dot{\hat{\theta}}_0 = -\gamma_0 e^T P B g C e, \\ \dot{\hat{\theta}}_i = -\gamma_i e^T P B \varphi_i(x_1, x_2, t), \end{cases} \tag{5.3.65}$$

and the matrix P satisfies Lyapunov inequality $PA_* + A_*^T P < 0$, where $A_* = A + B\theta_0 C$. Recall that all the variables needed for the adaptation algorithm can be measured if and only if $PB = C^T g$. Applying the lemma 5.1, we obtain that $\dot{V} < 0$ for $e \neq 0$ if and only if the adaptation algorithm has the form (5.3.59), (5.3.60) and the system $\dot{x} = Ax + Bu$, $y = gCx$ is hyper-minimum-phase, that holds by the theorem conditions. Therefore the function $V(t) = V(x(t), \hat{\theta}_0(t), \hat{\theta}(t))$ is bounded. Therefore, the functions $e(t)$, $\hat{\theta}_i(t)$ are

---

[4] Relative degree of a linear system is the difference between degrees of denominator and numerator polynomials of the system transfer function

also bounded in view of boundedness of $\varphi_i(x_1, x_2, t), i = 1, \ldots, N$. It follows from (5.3.65) that $\dot{V} = e^T(PA_* + A_*^T P)e \leq -\mu |e(t)|^2$ for some $\mu > 0$. Now we fall under conditions of the LaSalle theorem [380], where the function V is of the form (5.3.64). Since the boundedness of $V(t)$ implies boundedness of all the system states, it follows from the LaSalle theorem that $e(t) \to 0$ for $t \to \infty$.

To prove (5.3.63) first of all note that it follows from (5.3.54) and (5.3.59) that $\dot{\tilde{\theta}}(t) \to 0$ for $t \to \infty$. Differentiating (5.3.55) and taking into account boundedness of the functions $e, \tilde{\theta}, \varphi_d, \tilde{y}, \hat{\theta}_0$ and their time-derivatives we conclude that $\ddot{e}(t)$ is bounded. Applying the Barbalat lemma [157] yields $\dot{e}(t) \to 0$ for $t \to \infty$. Having in mind (5.3.59) we obtain that that $\tilde{\theta}(t)^T \varphi_d(t) \to 0$ for $t \to \infty$. Finally, (5.3.63) follows from the PE condition and Lemma 5.2. ∎

Remark 5.3. Theorem 5.1 actually provides necessary and sufficient conditions for existence of the Lyapunov function of the form (5.3.64) with the properties

$$\begin{cases} V(x, \hat{\theta}_0, \hat{\theta}, t) > 0 \text{ for } e \neq 0, \\ \dot{V}(x, \hat{\theta}_0, \hat{\theta}, t) < 0 \text{ for } e \neq 0. \end{cases} \quad (5.3.66)$$

It means that it is not possible to find another adaptation algorithm using any Lyapunov function of form (5.3.64) with properties (5.3.66). □

5.3.4 Synchronization and adaptive observers

In the previous section the problem of adaptive control of synchronization was studied. In a similar way one can solve the problem of adaptive observer-based synchronization [160]. Let us first formulate the problem, following [160]. Let the model of the uncontrolled system (transmitter) have the form

$$\dot{x}_d = Ax_d + \varphi_0(y_d) + B \sum_{i=1}^{N} \theta_i \varphi_i(y_d), \quad y_d = Cx_d, \quad (5.3.67)$$

where $x_d \in \mathbb{R}^n$ is the state vector of the transmitter; $y_d \in \mathbb{R}^l$ is the vector of outputs (transmitted signals); $\theta = \text{col}(\theta_1, \ldots, \theta_N)$ is the vector of the transmitter parameters. It is assumed that the nonlinearities $\varphi_i(\cdot), i = 0, 1, \ldots, N$, matrices A, C and vector B are known.

The problem is to design an adaptive observer (receiver) which is a dynamical system with input vector $y_d(t)$ and output vector $w(t) = \text{col}(\hat{x}(t), \hat{\theta})$, consisting of the vector of the transmitter state estimate $\hat{x}(t)$ and transmitter parameter estimate $\hat{\theta}$, ensuring the *observation goal* – convergence to zero of the estimation errors:

$$\lim_{t \to \infty} (\hat{x}(t) - x_d(t)) = 0, \quad (5.3.68)$$

$$\lim_{t \to \infty} \left(\hat{\theta}(t) - \theta\right) = 0. \quad (5.3.69)$$

Adaptive observer is designed in the form

$$\dot{\hat{x}} = A\hat{x} + \varphi_0(y_d) + B\Big(\sum_{i=1}^{N} \hat{\theta}_i \varphi_i(y_d) + \hat{\theta}_0 G(y_d - y)\Big),$$
$$y = Cx, \quad (5.3.70)$$

$$\dot{\hat{\theta}}_i = \psi_i(y_d, y), \quad i = 0, 1, \ldots, N, \quad (5.3.71)$$

where $x \in \mathbb{R}^n, y_d \in \mathbb{R}^l, \hat{\theta}_i \in \mathbb{R}$, and $G \in \mathbb{R}^l$ is the vector of the weighting coefficients. The adaptation algorithm (5.3.71) needs to be found. Though formally the observation problem differs from the control problem, the error equation can be transformed to the form (5.3.55), after introducing notations

$$e = x_d - \hat{x}, \quad \varphi_i = \varphi_i(y_d), \quad u = -\theta_0(y_d - C\hat{x}) + \sum_{i=1}^{N} \theta_i \varphi_i(y_d).$$

The adaptation algorithm designed by the speed-gradient method has the following form:

$$\dot{\hat{\theta}}_i = -\gamma_i (y - y_d) \varphi_i(y_d), \quad i = 1, \ldots, N, \quad (5.3.72)$$

$$\dot{\hat{\theta}}_0 = -\gamma_0 (y - y_d)^2, \quad (5.3.73)$$

where $\gamma_i > 0$.

Similarly to Theorem 5.1, the following statement providing conditions of adaptive synchronization can be proved.

Theorem 5.2 [160]. *Let the trajectories of the system (5.3.67) be bounded and the linear system with the transfer matrix $W(\lambda) = C(\lambda I - A)^{-1} B$ be hyper-minimum-phase.*

Then all the trajectories of the system (5.3.70), (5.3.72),(5.3.73) are bounded and the observation goal 5.3.68) holds. In addition, if the PE condition is fulfilled for the vector-function $(\varphi_1(y_d(t)), \ldots, \varphi_N(y_d(t)))$, then the estimation goal (5.3.69) holds: the adjustable parameters tend to their ideal values.

An important problem is taking into account the measurement noise. Let the received signal be $y_r(t) = y_d(t) + \xi(t)$, where $\xi(t)$ is channel noise. In presence of noise the algorithm (5.3.72), (5.3.73) may not guarantee synchronization and has to be regularized (robustified). One way of regularization is introducing the *parametric feedback*. It provides the following form of adaptive algorithm

$$\dot{\hat{\theta}}_i = -\gamma_i(y - y_r)\varphi_i(y_r) - \alpha_i\hat{\theta}_i, \quad i = 1, \ldots, m, \tag{5.3.74}$$

$$\dot{\hat{\theta}}_0 = -\gamma_0(y - y_r)^2 - \alpha_0\hat{\theta}_0, \tag{5.3.75}$$

where $\alpha_i > 0$ $(i = 0, 1, \ldots m)$ are regularization gains.

The following Theorem describes properties of the receiver with regularized adaptation algorithm (5.3.74), (5.3.75).

Theorem 5.3. Let the noise function $\xi(t)$ be bounded: $|\xi(t)| \leq \Delta_\xi$; all the trajectories of the transmitter 5.3.67 be bounded and linear system with transfer function $W(\lambda) = GC(\lambda I - A)^{-1}B$ be hyper-minimum-phase.

Then all the trajectories of the system (5.3.67), (5.3.70), (5.3.74), (5.3.75) are bounded and the goals

$$\overline{\lim}_{t\to\infty} (\hat{x}(t) - x_d(t)) \leq \Delta_x, \tag{5.3.76}$$

$$\overline{\lim}_{t\to\infty} \left(\hat{\theta}(t) - \theta\right) \leq \Delta \tag{5.3.77}$$

hold for some $\Delta > 0$, $\Delta_x > 0$. If, in addition, $\Delta_\xi > 0$ is sufficiently small, and gains $\alpha_i > 0$, $i = 0, 1, \ldots, m$ are chosen sufficiently small, then the values Δ_x in (5.3.76) and Δ in (5.3.77) can be chosen arbitrary small.

Theorem 5.3 follows from the results of [160] and [157].

In the works [25, 160] it was proposed to apply the adaptive observer for information transmission based on a chaotic carrier signal. The idea of the application is to encode the transmitted message by change (modulation) of the transmitter parameters $\theta_i, i = 1, \ldots, N$ and use the estimates of the parameters $\hat{\theta}_i, i = 1, \ldots, N$ on the receiver side to reconstruct the message. An advantage of such an adaptive receiver compared to that described in Example 5.7 is its potentially better robustness to the faults. Indeed, the breaks of synchronization caused by changes of the transmitter parameters (sending the message signal) lead to the corresponding changes in the parameters of the receiver and will not break synchronization. However, other faults (e.g., sudden faults in transmitter or in the communication channel) will lead to synchronization breaks and can be recognized by the receiver. Example of application of the above results to the information transmission is given in the next section. Further analysis of the transmission accuracy under bounded errors in the communication channel can be found in [23].

Finally, note that an important feasibility condition for the above adaptive synchronization schemes is the possibility of the error equation passification. By Lemma 5.1, passification needs fulfillment of the relative degree one condition: the relative degree of the linear part of the transmitter should be equal to one. In the paper [161] some new schemes of adaptive synchronization based on the notions of augmented error and high-order tuner allowing to weaken the relative degree condition are proposed and justified.

5.3.5 Example: Information transmission using chaotic Chua system

Consider a model information transmission system based on adaptive chaotic synchronization, where both transmitter and receiver system are implemented as Chua's circuits, following [25, 160].

Let the transmitter model be the Chua system (in dimensionless form) [116]

$$\begin{aligned}\dot{x}_{d_1} &= p(x_{d_2} - x_{d_1} + f(x_{d_1}) + sf_1(x_{d_1})) \\ \dot{x}_{d_2} &= x_{d_1} - x_{d_2} + x_{d3} \\ \dot{x}_{d_3} &= -qx_{d_2}\end{aligned} \quad (5.3.78)$$

where $f(z) = M_0 z + 0.5(M_1 - M_0)f_1(z)$, $f_1(z) = |z+1| - |z-1|$, M_0, M_1, p, q are the transmitter parameters, $s = s(t)$ is the signal to be reconstructed in the receiver. Assume that the transmitted signal is $y_r(t) = x_{d_1}(t)$, and the values of the parameters p, q are known. It is well known that for some value of the system parameters the system (5.3.78) may exhibit a chaotic behavior.

The parameters M_0, M_1 are assumed to be *a priori* unknown which motivates the use of an adaptation for the receiver design. The receiver designed according to the results of the previous section is modeled as

$$\begin{aligned}\dot{x}_1 &= p(x_2 - x_1 + f(y_r) + c_1 f_1(y_r) + c_0(x_1 - y_r)), \\ \dot{x}_2 &= x_1 - x_2 + x_3, \\ \dot{x}_3 &= -qx_2,\end{aligned} \quad (5.3.79)$$

where c_0, c_1 are the adjustable parameters. The adaptation algorithm (5.3.74), (5.3.75), takes the form

$$\begin{aligned}\dot{c}_0 &= -\gamma_0(y_r - x_1)^2 - \alpha_0 c_0, \\ \dot{c}_1 &= -\gamma_1(x_1 - y_r)f_1(y_r) - \alpha_1 c_1,\end{aligned} \quad (5.3.80)$$

where γ_0, γ_1 are the adaptation gains, $\alpha_0 \geq 0, \alpha_1 \geq 0$ are the regularization gains.

First, we examine the ability of the system (5.3.79), (5.3.80) to receive and to decode messages without noise. To this end we verify the conditions of the Theorem 5.2 assuming that $s(t) = $ const and $\alpha_0 = \alpha_1 = 0$. Clearly, if $s(t)$ is a time-varying binary signal, we can only expect that the results of Theorem 5.2 can be used if the parameter estimation is fast enough, at least much faster than the actual parameter modulation. Writing the error equations yields

$$\begin{cases}\dot{e}_1 = p(e_2 - e_1 + (c_1 - s)f_1(y_r) + c_0 e_1) \\ \dot{e}_2 = e_1 - e_2 + e_3 \\ \dot{e}_3 = -qe_2,\end{cases} \quad (5.3.81)$$

where $e_i = x_i - x_{d_i}$, $i = 1, 2, 3$. The system (5.3.81) is, obviously in Lur'e form with

5 Controlled Synchronization

$$A = \begin{bmatrix} -p & p & 0 \\ 1 & -1 & 1 \\ 0 & -q & 0 \end{bmatrix}, \quad B = \begin{bmatrix} 1 \\ 0 \\ 0 \end{bmatrix}, \quad C = [1\ 0\ 0],$$

$\hat{\theta}_1 = c_1$, $\theta_1 = s$, $\theta_0 = c_0$.

The transfer function of the linear part is

$$W(\lambda) = \frac{\lambda^2 + \lambda + q}{\lambda^3 + (p+1)\lambda^2 + q\lambda + pq} \tag{5.3.82}$$

We see that the order of the system is $n = 3$, while the numerator polynomial is Hurwitz and has degree 2 for all $q > 0$ and all real p. Therefore, the hyper-minimum-phase condition holds for $q > 0$ and any p, M_0, M_1. Thus, Theorem 5.2 yields the boundedness of all receiver trajectories $x(t)$ and convergence of the observation error: $e(t) \to 0$. In particular, $y_r(t) - x_1(t) \to 0$. Furthermore, to be able to reconstruct the signal $s(t)$ the receiver should provide convergence $c_1(t) - s \to 0$ for constant s. According to Theorem 5.2, this will be the case if the PE condition (see Definition 5.4) holds, which reads as

$$\int_{t_0}^{t_0+T} f_1^2(y_r(t))\, dt \geq \varrho \tag{5.3.83}$$

for some $T > 0$, $\varrho > 0$ and all $t_0 \geq 0$. To verify (5.3.83), we note that condition (5.3.83) basically means that the trajectory of the transmitter $x_d(t)$ does not converge to the plane $x_{d_1} = 0$ when $t \to \infty$. This is not the case, at least when the system (5.3.78) exhibits chaotic behavior. Indeed, in this case the value $x_{d_1}(t)$ leaves the interval $(-1, 1)$ (where $f_1(z)$ is linear) infinitely many times, say at t_k, $k = 1, 2, \ldots$. The time intervals $\Delta t_k = t_{k+1} - t_k$ between t_k can be overbounded by constant, if the trajectory does not converge to the set $x_{d_1} = 0$.

We may also evaluate a lower bound for ϱ in (5.3.83):

$$\varrho_0 = \underline{\lim}_{T \to \infty} \frac{1}{T} \int_0^T f_1^2(x_{d_1}(t))\, dt. \tag{5.3.84}$$

The value of ϱ_0 characterizes the parameter convergence rate. It follows from the adaptive control theory results (see, e.g., [388]) that if $\varrho_0 > 0$, then the convergence $c_1(t) - s \to 0$ is exponential, with rate $\gamma_1 \varrho_0$, at least for sufficiently small $\gamma_1 > 0$. Ergodicity arguments suggest that

$$\varrho_0 \geq \frac{\overline{x}_{d_1}^2}{\mu}, \tag{5.3.85}$$

where $\overline{x}_{d_1}^2$ is the average value of $x_{d_1}^2(t)$ over the attractor Ω, and $\mu = \sup_{x \in \Omega} |x_{d_1}(t)|$.

Let the channel be subjected to white noise $\xi(t)$ added to the transmitter output, so that the received signal $y_r(t)$ is modeled as

$$y_r(t) = y_d(t) + \xi(t), \tag{5.3.86}$$

where $\xi(t)$ is a Gaussian white noise with zero average value and intensity σ.[5] In the noisy case the robustified adaptation algorithm with $\alpha_0 > 0, \alpha_1 > 0$ should be applied. Boundedness of all errors follows from Theorem 5.3.

Theoretical analysis provided by Theorems 5.2, 5.3 is usually not sufficient for practice. In order to evaluate the system performance computer simulations should be performed. Simulation results for the above scheme are shown below. Parameter values are selected as $p = 9; q = 14.286; M_0 = 5/7; M_1 = -6/7$. For these parameter values the system (5.3.78) possesses a chaotic attractor, see Fig. 5.3.1 resembling that of the system used in [116] (after some rescaling of space and time variables).

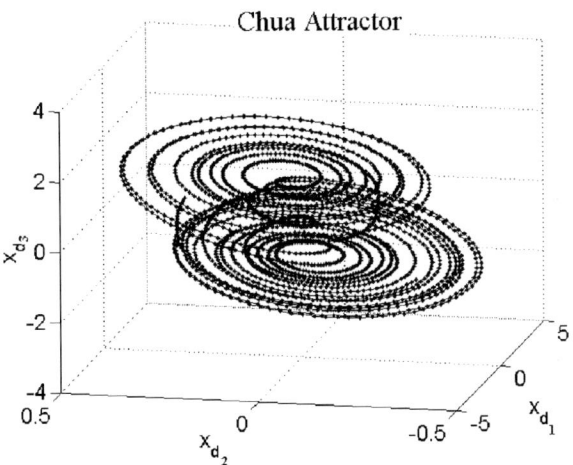

Fig. 5.3.1. Double scroll attractor of Chua curcuit.

The initial conditions for the transmitter were taken as $x_d(0) = [0.3 \ 0.3 \ 0.3]$. For the receiver zero initial conditions were chosen for the state x_0 as well as for the adjustable parameters $c_0(0), c_1(0)$. In order to eliminate the influence of initial conditions no message was transmitted during the first 20 time units ("tuning" or "calibration" of the receiver), i.e., $s(t) \equiv 1$ for $0 \leq t \leq 20$. The time history of observation errors and parameter estimates during tuning show that all observation errors and parameter estimation error $c_1(t) - s$ tend to zero rapidly. The value $c_0(t)$ tends to some constant value.

---

[5] More precisely, $\xi(t)$ is modeled as a piecewise constant random process with sample time Δ_t and $\xi(t_k) = \zeta_k \sqrt{\Delta_t}$, $(k = 0, 1, 2, \ldots, t_k = k\Delta_t)$, where ζ_k are Gaussian random numbers, having zero mean and the standard deviation σ.

After the tuning period the square wave message

$$s(t) = s_0 + s_1 \operatorname{sign} \sin\left(\frac{2\pi t}{T_0}\right), \qquad (5.3.87)$$

where $s_0 = 1.005$, $s_2 = 0.005$ was sent. Received signal $y_r(t)$ is shown on the Fig. 5.3.2.

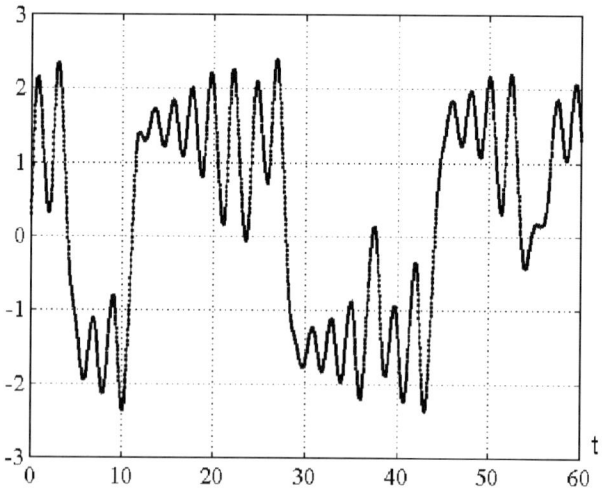

Fig. 5.3.2. Received signal $y_r(t)$.

The simulation shows, that the reconstructed signal $y(t)$ coincides with the transmitted signal $y_r(t)$ with very good accuracy. However, both observation errors and estimation errors do not decay completely during the interval when $s(t)$ is constant. Nevertheless, a reliable reconstruction of the signal $s(t)$ is very well possible. The accuracy of estimation can be easily improved by increasing the adaptation gain γ_1. The achievable information transmission rate depends on the highest frequencies in the carrier spectrum. Fig. 5.3.3 a shows the message signal $s(t)$ (with the period $T_0 = 10$), its estimate via adaptive observer algorithm (5.3.79), (5.3.80) $\hat{s}(t)$ and output of the first-order low-pass filter $s_f(t)$. This filter is used for separation the message in the case of the noisy channel. In the simulation were taken following parameters of the algorithm: $\gamma_0 = 10$, $\alpha_0 = 1$ $\gamma_1 = 5$, $\alpha_1 = 0.2$, filter pass band is equal to 3.5. Fig. 5.3.3 b illustrates the influence of the noise with $\sigma = 10^{-3}$ in the channel. One can notice, that the message can be recognized in this case too.

5.4 Synchronization of two coupled pendulums

Consider the special case of the diffusively coupled oscillator model, namely, two pendulums connected with the torsion spring:

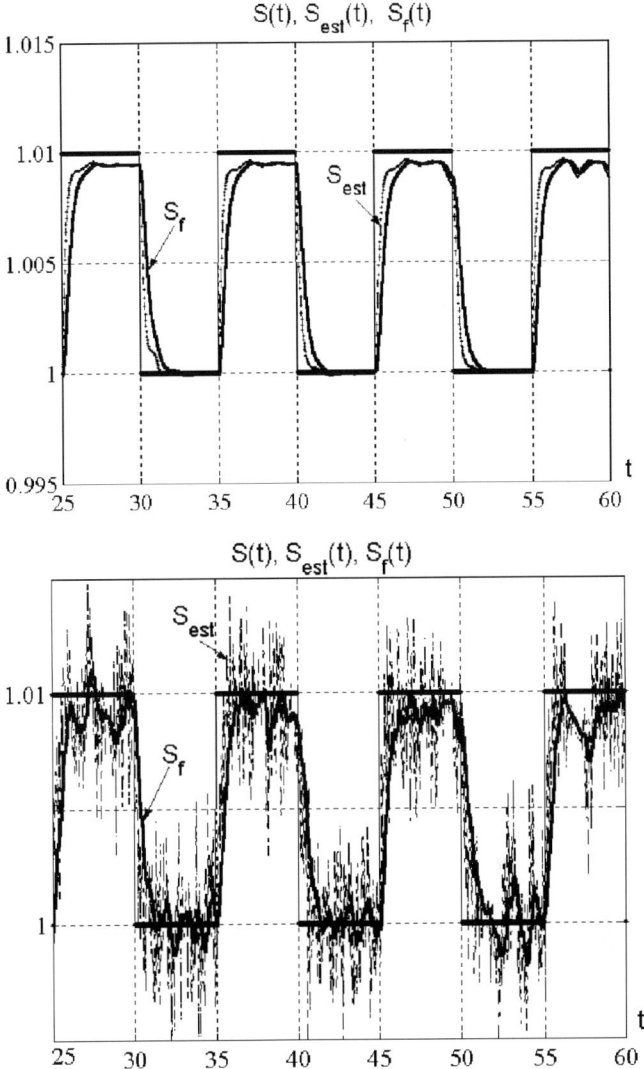

Fig. 5.3.3. Parameter estimation by means of the adaptive observer (5.3.79), (5.3.80).

$$\begin{cases} \ddot{\varphi}_1 + \rho\dot{\varphi}_1 + \omega^2 \sin\varphi_1 + k(\varphi_1 - \varphi_2) = u(t), \\ \ddot{\varphi}_2 + \rho\dot{\varphi}_2 + \omega^2 \sin\varphi_2 + k(\varphi_2 - \varphi_1) = 0, \end{cases} \quad (5.4.88)$$

where $\varphi_i(t)$ are the rotation angles of pendulums ($i = 1, 2$); $u(t)$ is the external torque (control action) applied to the first pendulum; ω, k, ρ are the system parameters: ω is the natural frequency of small oscillations, k is the coupling parameter (e.g., stiffness of the string), ρ is the viscous friction gain.

The total energy of the system (5.4.88) $H(x)$ can be written as follows

$$H(x) = \frac{1}{2}\dot{\varphi}_1^2 + \omega^2(1 - \cos\varphi_1) + \frac{1}{2}\dot{\varphi}_2^2 \\ + \omega^2(1 - \cos\varphi_2) + \frac{k}{2}(\varphi_1 - \varphi_2)^2 \quad (5.4.89)$$

where $x(t) \in \mathbb{R}^4$ stands for the state vector $x(t) = (\varphi_1, \varphi_2, \dot{\varphi}_1, \dot{\varphi}_2)^T$.

Consider the problem of excitation of synchronous oscillations with the desired amplitude by means of small feedback. The problem can be understood as achieving the given energy level of the system with the additional requirement that pendulums should have either coincident or opposite phases of oscillation.

In order to apply the speed-gradient procedure of Section 2.4, let us introduce two auxiliary goal functions as follows

$$Q_\varphi(\dot{\varphi}_1, \dot{\varphi}_2) = \frac{1}{2}(\delta_\varphi)^2, \\ Q_H(x) = \frac{1}{2}(H(x) - H_*)^2, \quad (5.4.90)$$

where $\delta_\varphi = \dot{\varphi}_1 + \sigma\dot{\varphi}_2$, $\sigma \in \{-1, 1\}$; H_* is the prescribed value of the total energy. The minimum value of the function Q_φ meets the "coincident/opposite phases" requirement (at least for small initial phases $\varphi_1(0), \varphi_2(0)$) : $Q_\varphi(\dot{\varphi}_1, \dot{\varphi}_2) \equiv 0$ if and only if $\dot{\varphi}_1 \equiv -\sigma\dot{\varphi}_2$. Hence option $\sigma = 1$ sets *anti phase* desired pendulums oscillations, while $\sigma = -1$ sets *in phase* oscillations. The minimization of Q_H means achievement of the desired amplitude of the oscillations.

In order to design the control algorithm, the weighted objective function $Q(x)$ is introduced as the weighted sum of Q_φ and Q_H, namely

$$Q(x) = \alpha Q_\varphi(\dot{\varphi}_1, \dot{\varphi}_2) + (1 - \alpha)Q_H(x), \quad (5.4.91)$$

where α, $0 \leq \alpha \leq 1$ is a given weighting coefficient.

According to the speed-gradient procedure of Section 2.4, the following control law is obtained:

$$u(t) = -\gamma\left(\alpha\delta_\varphi(t) + (1-\alpha)\delta_H(t)\dot{\varphi}_1(t)\right), \\ \delta_\varphi(t) = \dot{\varphi}_1(t) + \sigma\dot{\varphi}_2(t), \\ \delta_H(t) = H(x(t)) - H_*, \quad (5.4.92)$$

where $\sigma \in \{-1, 1\}$ is a phase sign parameter; α is a weighting coefficient; $\gamma > 0$ is a gain coefficient. The existing analytical results, see Chapters 3, 4 and [164] do not apply to this problem. Therefore the problem of finding sufficient conditions for the achievement of the control goal $Q(x(t)) \to 0$ remains open.

By analogy with (4.2.41), the following *relay speed gradient* control law can be written:

$$u(t) = -\gamma\,\text{sign}\left(\alpha\delta_\varphi(t) + (1-\alpha)\delta_H(t)\dot{\varphi}_1(t)\right). \quad (5.4.93)$$

5.4 Synchronization of two coupled pendulums

The control law (5.4.92) has been proposed and numerically examined in [26].

Let us study the properties of the system with the algorithm (5.4.93). At the first stage let us assume $\alpha = 0$, i.e., consider the closed-loop system with a pure *energy-control* algorithm

$$u = -\gamma \operatorname{sign}(H - H_*) \operatorname{sign} \dot{\varphi}_1(t), \quad (5.4.94)$$

where the total energy H is given by (5.4.89), H_* is a desired value of H. Note that the control law (5.4.94) has the same form as the law (4.2.41).

The question under consideration is: if no requirement on the phase shift is given (i.e., it is taken $\alpha = 0$ in (5.4.93)), what will be the phase shift in the steady-state oscillation mode? To answer that question the system (5.4.88), (5.4.94) has been numerically studied by means of computer simulations. The following parameter values and the initial conditions were chosen: $\omega^2 = 10 \, \text{s}^{-2}$, $\rho = 0.1 \, \text{s}^{-1}$, $\gamma = 1 \, \text{s}$, $H_* = 20 \, \text{s}^{-2}$, $\varphi_1(0) = \pi/2$, $\dot{\varphi}_1(0) = \dot{\varphi}_2(0) = 0$. Initial value $\varphi_2(0)$ was varied in the segment $[-3/4\pi, 3/4\pi]$, the coupling coefficient k changed from 0.1 to $2 \, \text{s}^{-2}$. The simulation time was equal to 450 s; the fixed-step Dormand–Prince method with a step 0.025 s was used. Some results are shown in Fig. 5.4.4. In Fig. 5.4.4 the plots of the *frequency of oscillation* Ω and the *phase shift* $\Delta \psi$ in the steady-state mode versus the coupling parameter k and the initial condition $\varphi_2(0)$ are pictured. It is shown that in some domain of the plane $(k, \varphi_2(0))$ the steady phase shift is about zero, i.e., the pendulums fall into in phase synchronous oscillations, while in the complement domain the phase shift is about π – the motions are anti phase. Note, that the oscillation frequencies for anti phase motion exceed those ones for the in phase motion.

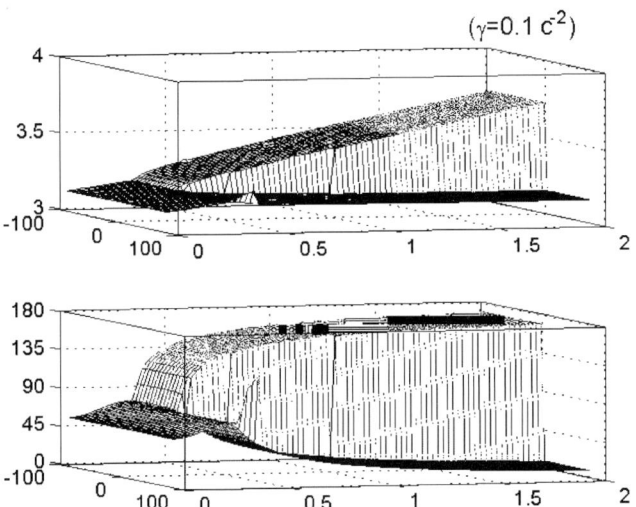

Fig. 5.4.4. Oscillations frequency Ω_0 and the phase shift $\Delta \psi$ versus k, $\varphi_2(0)$. The control law (5.4.94), $\gamma = 1$ s.

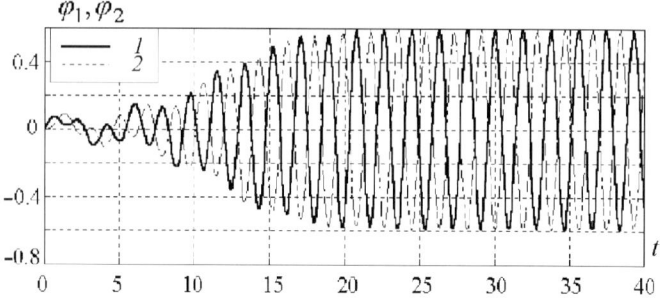

Fig. 5.4.5. Excitation of anti phase oscillations of coupled pendulums.

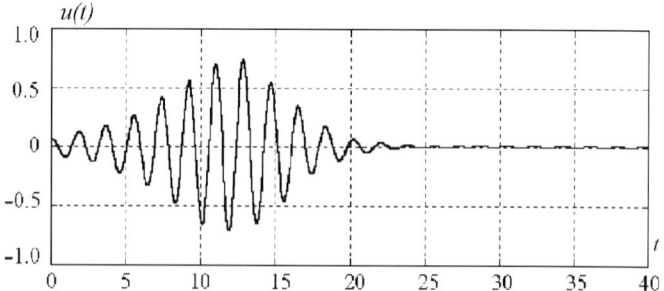

Fig. 5.4.6. Controlling signal $u(t)$.

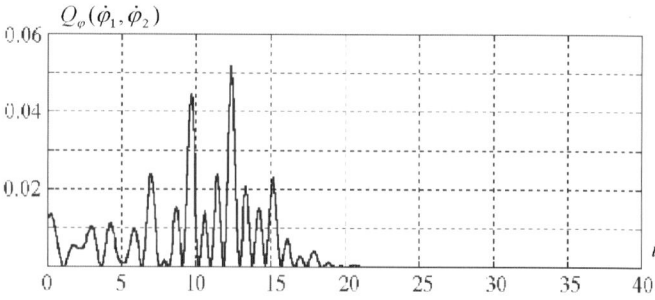

Fig. 5.4.7. Time history of the synchronization goal function $Q_\varphi(\varphi(t)_1, \varphi(t)_2)$.

This effect has a clear physical explanation: in the anti phase mode the spring torque is added to the torque of a gravitational force.

Let us consider now the general form (5.4.93). The simulation results for $\alpha = 0.7$ are shown in Figs. 5.4.5–5.4.9. Two cases of the damping parameter ρ were considered: $\rho = 0$ (the lossless case), and $\rho = 0.1$ s^{-1}. The following initial conditions were chosen: $\varphi_1(0) = \varphi_2(0) = 0$, $\dot\varphi_1(0) = \dot\varphi_2(0) = 10^{-5}$ s^{-1}.

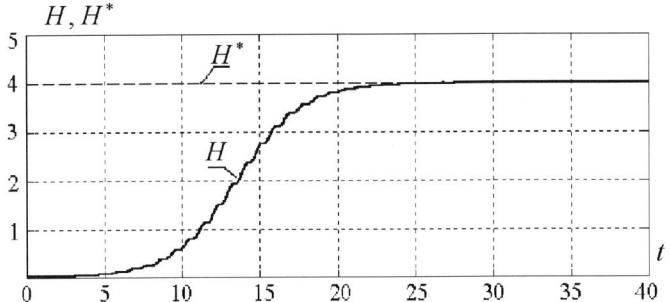

Fig. 5.4.8. Time history of the oscillations energy H_t.

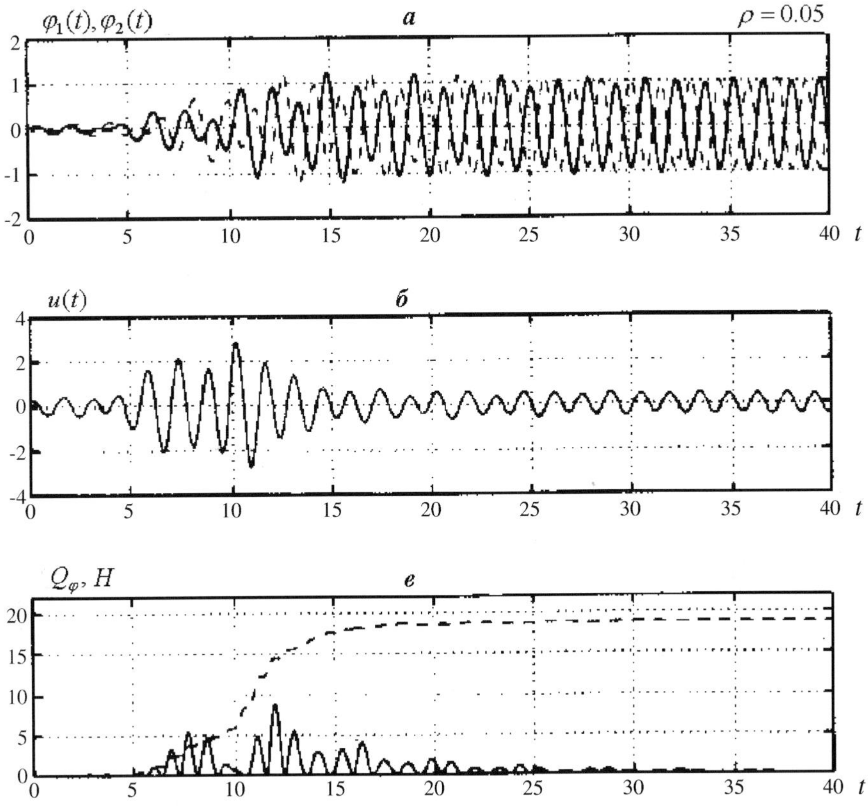

Fig. 5.4.9. Time history of synchronization in presence of dissipation.

It is seen that both pendulums fall in anti phase oscillations if $\sigma = 1$ and in phase oscillations if $\sigma = -1$. The relation between transient times for H and for Q_φ can be changed by means of changing the weight coefficient α. In the lossless case the control amplitude can be arbitrarily decreased by decreasing the gain γ.

6
Control of Chaos

The field related to control of chaotic systems was rapidly developing during the 1990s. Its state-of-the-art in the beginning of the 21st century is presented in this chapter, following [27, 28, 145, 153]. Necessary preliminary material is given related to notion and properties of chaotic systems, models of the controlled plants and control goals. Several major branches of research are discussed in detail:

– feedforward or "nonfeedback" control (based on periodic excitation of the system);
– OGY method (based on linearization of Poincaré map);
– Pyragas method (based on time-delay feedback);
– traditional control engineering methods of linear, nonlinear, and adaptive control.

Some unsolved problems concerning the justification of chaos control methods are presented. Other directions of research are outlined such as chaotic mixing, generation of chaos (chaotization), etc. Areas of existing and potential applications in science and engineering are pointed out.

6.1 Introduction

Chaotic system is a deterministic dynamical system exhibiting irregular, seemingly random behavior. Two trajectories of a chaotic system starting close to each other will diverge after some time (such an unstable behavior is often called "sensitive dependence on initial conditions"). Mathematically, chaotic systems are characterized by local instability and global boundedness of the trajectories. Since local instability of a linear system implies unboundedness (infinite growth) of its solutions, chaotic system should be necessarily nonlinear, i.e., should be described by a nonlinear mathematical model.

Control of chaos, or control of chaotic systems is the boundary field between control theory and dynamical systems theory studying when and how it is possible to control systems exhibiting irregular, chaotic behavior. Control of chaos is closely related to nonlinear control, many methods of nonlinear control are applicable to chaotic systems. However, control of chaotic systems has some specific features.

A key property of chaotic systems is its instability: sensitive dependence on initial conditions. An important consequence is high sensitivity with respect to changes of input (controlling action). It means that small changes of control may produce large variations in systems behavior. Such a phenomenon and its implications in physics were described in the seminal paper [331] that triggered an explosion of activities and thousands of publications.

A typical control goal when controlling chaotic systems is to transform a chaotic trajectory into a periodic one. In terms of control theory it means stabilization of an unstable periodic orbit or equilibrium. A specific feature of this problem is the possibility of achieving the goal by means of an arbitrarily small control action. Other control goals like synchronization and chaotization can also be achieved by small control in many cases.

Since 1975, when the term "chaos" was coined by Li and Yorke [267], chaotic phenomena and chaotic behavior have been observed in numerous natural and model systems in physics, chemistry, biology, ecology, etc. Paradigm of chaos allows to better understand inherent properties of natural systems. Engineering applications are rapidly developing in areas such as lasers and plasma technologies, mechanical and chemical engineering, and telecommunications. Possibilities of controlling complex behavior by means of small control open new horizons both in science and in technology.

Development of new methods for control of chaos or "control by tiny corrections" [103, 164, 403] may be of utmost importance for sustained development of humanity. They may be efficient for solving problems where applying stronger control is not possible either because of lack of resources (like in many large scale systems: economies, energy systems, weather control, etc.) or because intervening natural dynamics is undesirable (e.g., in biological and biomedical applications, ecology systems).

It is worth noticing that, in spite of the enormous number of published papers, not many rigorous mathematical results are so far available. A great deal of results are justified by computer simulations rather than by analytical tools and many problems remain unsolved. Main approaches to controlling chaotic behavior are described below. Before exposition of the methods some mathematical preliminaries are given concerning system models, control goals, and properties of chaotic systems.

6.2 Notion of chaos

6.2.1 Definitions of chaos

There exist different definitions of chaotic system and chaotic behavior. The following quotation gives an idea of the situation.

6.2 Notion of chaos

There are many possible definitions of chaos. In fact, there is no general agreement within the scientific community as to what constitutes a chaotic dynamical system. (R. Devaney. A first course in chaotic dynamical systems. Addison-Wesley, 1992.)

In most of definitions chaotic processes are treated as solutions of nonlinear differential or difference equations, characterized by local instability and global boundedness. It means that solutions with close initial conditions will diverge to some finite distance after some time (so called "sensitive dependence on initial conditions"). Below a typical definition and a typical criterion of chaos are introduced. More detail and discussions see e.g. in [223, 304, 312, 332, 419].

Consider the system
$$\dot{x} = f(x), \tag{6.2.1}$$
where $x \in \mathbb{R}^n$ is n-dimensional state vector, $\dot{x} = d/dt$ stands for the time derivative of x.

To define chaotic system the notions of attracting set, attractor, and a chaotic attractor are used.

Definition 6.1. A set B_0 is called the *attracting set* for the system (6.2.1) if there exists an open set B such that $B_0 \subset B$ and
$$\lim_{t \to \infty} \text{dist}(x(t), B_0) = 0 \tag{6.2.2}$$
for any solution $x(t)$ of (6.2.1) with $x(0) \in B$.

Definition 6.2. A closed attracting set B_0 is called the *attractor* if it is minimal, i.e., there is no smaller attracting subset of B_0. The set of initial conditions B for which (6.2.2) holds is called *basin of attraction*.

The minimality property of attractor expresses the fact that the trajectory will pass through any vicinity of any point of attractor. Such a property is also known as *transitivity*.

Definition 6.3. An attractor B_0 is called *chaotic* if it is bounded and all the trajectories starting from it are Lyapunov unstable. The system (6.2.1) is called *chaotic* if it possesses at least one chaotic attractor.

Although the above definition is used very often, some alternatives have been suggested in the literature. For example, the definition due to R.L. Devaney [118] requires additionally that periodic trajectories are dense in the attractor. Note that this requirement is often redundant because it follows from the recurrence property inherent for both periodic and chaotic trajectories (see the Anosov lemma below). Some authors prefer to use the term "strange attractor" instead of "chaotic attractor." The term "strange attractor" introduced by D. Ruelle and F. Takens [383] in 1971 means that the attractor is a porous (fractal) set that cannot be represented as a piece of manifold and therefore has a noninteger dimension (see the discussion of fractal dimension below). Though the concepts of "strange attractor" and "chaotic attractor" are indistinguishable in many applications, there exist also strange nonchaotic attractors proven theoretically and experimentally.

An important property of chaotic trajectories for control purposes is *recurrence*: they return to any vicinity of any past value. A general definition of recurrence is as follows.

Definition 6.4. The function $x : \mathbb{R}^1 \to \mathbb{R}^n$ is called *recurrent* if for any $\varepsilon > 0$ there exists $T_\varepsilon > 0$ such that for any $t \geq 0$ there exists $T(t,\varepsilon), 0 < T(t,\epsilon) < T_\varepsilon$ such that $|x(t + T(t,\varepsilon)) - x(t)| < \varepsilon$.

Introduction of recurrence property can be traced back to the end of the 19th and the beginning of the 20th century and related to the names of H. Poincaré and G. Birkhoff. The recurrent function necessarily returns to any vicinity of any previous value at least once and, therefore, infinitely many times. Although time intervals between the returns are not equal, they are bounded, i.e., cannot grow infinitely with time.

Recurrent trajectories possess two important properties formulated in Pugh lemma and Anosov lemma in the 1960s and providing formal support of the claim that chaotic attractor is the closure of all the periodic trajectories contained in it, see [226].

Lemma 6.1. (C.C. Pugh). Let $\bar{x}(t), t \geq 0$ be the recurrent trajectory of the system (6.2.1) with smooth $f(x)$. Then for any $\varepsilon > 0$ there exists smooth function $f_1(x)$ such that $||f_1(x)||_\infty + ||Df_1(x)||_\infty < \varepsilon$ and the solution $x(t)$ of the system $\dot{x} = f(x) + f_1(x)$ with same initial condition $x(0) = \bar{x}(0)$ is periodic.

Lemma 6.2. (D.V. Anosov). Let $\bar{x}(t), t \geq 0$ be the recurrent trajectory of the system (6.2.1) with smooth $f(x)$. Then for any $\varepsilon > 0$ there exists x^* such that $||x^* - \bar{x}(0)|| < \varepsilon$ and the solution $x(t)$ of the system (6.2.1) with initial condition $x(0) = x^*$ is periodic.

Lemma 6.1 states that a recurrent trajectory can be transformed into a periodic one by means of arbitrarily small change of the right-hand side of the differential equation. Lemma 6.2 states that a recurrent trajectory can be transformed into a periodic one by means of arbitrarily small change of initial conditions.

The notion of attractor is related to the criterion of recurrence formulated by G. Birkhoff in 1927.

Theorem 6.1. (G. Birkhoff). *Any trajectory contained in the compact minimal invariant set is recurrent. And any compact minimal invariant set is the closure of some recurrent trajectory.*

Recall that the point which is the limit point of a trajectory for $t \to +\infty$ is termed *ω-limit point* of this trajectory [226]. It follows from the Birkhoff theorem that any solution starting from its ω-limit set (the set of all ω-limit points) is recurrent. Under additional assumption that ω-limit set of $\bar{x}(t)$ is attractor, any chaotic trajectory, starting from its ω-limit set is recurrent.

In other words, a steady-state behavior of a dynamical system is typically recurrent. On the contrary, the transient behavior cannot be recurrent.

6.2.2 Criteria of chaos

Since formal verification of chaoticity of a system behavior is usually very difficult, various numerical criteria are used. The most standard criterion of chaotic behavior is based on computation of the largest Lyapunov exponent. For a linear system

$$\dot{x} = A(t)x \qquad (6.2.3)$$

the largest Lyapunov exponent is defined as follows

$$\varrho_L = \overline{\lim_{t \to \infty}} \frac{\ln|\Phi(t,t_0)|}{t - t_0}, \qquad (6.2.4)$$

where $\Phi(t,\tau)$ is the transition matrix of the system (6.2.3) (square matrix, generating solutions of (6.2.3) as $x(t) = \Phi(t,\tau)x(\tau)$ for all $t,\tau \in \mathbb{R}^1$. For a nonlinear system the largest Lyapunov exponent along a solution $\bar{x}(t)$ is defined as the largest Lyapunov exponent of the system (6.2.1), linearized along $\bar{x}(t)$ (i.e., $A(t) = \partial f(\bar{x}(t))/\partial x$)

If the trajectory $\bar{x}(t)$ of (6.2.1) is bounded and $\varrho_L > 0$, then $\bar{x}(t)$ is chaotic. Note that it is only sufficient condition. Necessary conditions are formulated using more subtle characteristics: the so-called Bohl exponents. The value of $\varrho_L > 0$ indicates the degree of exponential instability of the system.

To verify chaoticity of discrete-time systems the Sharkovsky–Li–Yorke criterion and Marotto theorem can be used. At first we formulate a simple and beautiful criterion of chaos for one-dimensional maps established in 1975 by T. Li and J. Yorke [267].

Theorem 6.2. (Li-Yorke). *If F is a continuous map of a segment of the real axis to itself, and F has a periodic point of period 3, then F is chaotic (in the sense of Definition 6.3).*

The periodicity result of Li and Yorke can be derived from a more general theorem established in 1964 by A. Sharkovsky [400]. Introduce the following ordering in the set of natural numbers:

$$3 \succ 5 \succ \ldots \succ 3 \cdot 2 \succ 5 \cdot 2 \succ \ldots \succ 3 \cdot 2^2 \succ 5 \cdot 2^2 \succ \ldots \succ 2^n \succ \ldots \succ 2^2 \succ 2 \succ 1$$

Theorem 6.3. (Sharkovsky). *Consider a continuous map $F : \mathbb{R}^1 \to \mathbb{R}^1$. If F has a periodic point of period k, then F has a periodic point of any period n, where $k \succ n$.*

There are no direct analogs of the above results for multidimensional systems. One sufficient condition of chaos was suggested by F. Marotto in 1978 [286].

Theorem 6.4. (Marotto). *Let $F : \mathbb{R}^n \to \mathbb{R}^n$ be a smooth map and x be its fixed point. Suppose that for some $\varepsilon > 0$ and for some natural number n there exists y such that $\|y - x\| < \varepsilon$, $F^n y = x$, $\det(DF^n y) \neq 0$ and $\|Fx_1 - Fx_2\| > \|y_1 - y_2\|$ for all y_1, y_2, satisfying $\|y_1 - x\| < \varepsilon$, $\|y_2 - x\| < \varepsilon$. Then F is chaotic in the sense of Definition 6.3.*

6.2.3 Delayed coordinates and Poincaré map

Two other important tools to work with chaotic systems are delayed coordinates and Poincaré map.

Delayed coordinates. Let only a scalar output $y(t) = h(x(t))$ of the system (6.2.1) be available for measurement. It is often convenient to rewrite the model of a nonlinear system in terms of measured output variables. To this end the vector of delayed coordinates is defined as

$$X(t) = [y(t), y(t - \tau), \ldots, y(t - (N-1)\tau)]^\mathrm{T} \in \mathbb{R}^N.$$

The initial system model (6.2.1) can be transformed in the delayed coordinates as $\dot{X} = \bar{F}(X(t))$. Embedding theorems claim that if $N > 2n$, where n is dimension of the initial system (6.2.1) then generically there exists a diffeomorphism (smooth and smoothly invertible mapping) between the state space of initial system and a subspace of the state space of the transformed system such that if the initial system has an attractor of some dimension then the transformed system will have an attractor of the same dimension.

Poincaré map. Poincaré map allows to consider discrete-time system instead of continuous-time one and to reduce by 1 its dimension. To define Poincaré map assume that $\bar{x}(t)$ be T-periodic solution (6.2.1) starting from x_0, i.e., $\bar{x}(t + T) = x(t)$ for all $t \geq t_0, x(t_0) = x_0$. Let S be a smooth surface (*transverse surface* or *cross-section* across x_0), defined by the equation $s(x) = 0$ where $s : \mathbb{R}^n \to \mathbb{R}^1$ is a smooth scalar function such that it intersects the trajectory in x_0 transversely, i.e., $s(x_0) = 0, \nabla s(x_0)^\mathrm{T} F(x) \neq 0$. It can be shown that the solution starting from $x \in S = \{x : s(x) = 0\}$ close to x_0 will cross the surface $s(x) = 0$ again at least once. Let $\tau = \tau(x)$ be the time of the first return and $x(\tau) \in S$ be the point of the first return.

Definition 6.5. *The mapping $P : x \mapsto x(\tau)$ is called the Poincaré map or return map.*

Since S lies in a $n - 1$-dimensional manifold, where n is dimension of $x(t)$, a coordinate chart can be introduced in S and after a coordinate change the discretized system model can be written as follows:

$$x_{k+1} = P(x_k), k = 0, 1, 2, \ldots,$$

where x_k are $n - 1$-dimensional vectors. Behaviors of the initial continuous system and of the discretized one are qualitatively the same. It motivates study of the system dynamics by means of Poincaré map.

For controlled systems described by models $\dot{x} = F(x, u)$, instead of (6.2.1) Poincaré section itself may depend on the control variable u and therefore Poincaré map also depends on control: $P : (x, u) \mapsto x(\tau)$. In this case the map, called *controlled Poincaré map* is formally defined and studied in [164].

6.3 Models of controlled systems and control goals

Let us recall material of Sections 2.1 and 2.2, focusing on models and goals, typical for control of chaos.

Models of controlled systems. A formal statement of a control problem typically begins with a model of the system to be controlled (*controlled system* or *controlled plant*) and a model of the control objective (*control goal*). If the plant model is not given *a priori* (as in many real life applications) some approximate model should be determined in some way. Several classes of models are considered in the literature related to control of chaos. The most common class consists of continuous systems with lumped parameters described in state space by differential equations

$$\dot{x} = F(x, u), \qquad (6.3.5)$$

where x is n-dimensional vector of the state variables; u is m-dimensional vector of inputs (control variables). The vector-function $F(x, u)$ is usually assumed continuously differentiable which guarantees local existence and uniqueness of solutions of (6.3.5). The model should also include the description of measurements, i.e., the l-dimensional vector of output variables y should be defined, for example

$$y = h(x). \qquad (6.3.6)$$

If the outputs are not defined explicitly, it is assumed that all the state variables are available for measurement, i.e., $y = x$.

More detail concerning different classes of controlled system models and their peculiarities can be found in Chapter 2.

Control goals: Stabilization. A typical goal for control of chaotic systems is stabilization of an unstable periodic solution (orbit). Let $x_*(t)$ be the T-periodic solution of the free ($u(t) = 0$) system (6.3.5) with initial condition $x_*(0) = x_{*0}$, i.e., $x_*(t + T) = x_*(t)$ for all $t \geq 0$. If the solution $x_*(t)$ is unstable, a reasonable goal is stabilization or driving solutions $x(t)$ of (6.3.5) to $x_*(t)$ in the sense of fulfillment of the limit relation

$$\lim_{t \to \infty} [x(t) - x_*(t)] = 0 \qquad (6.3.7)$$

or driving the output $y(t)$ to the desired output function $y_*(t)$, i.e.,

$$\lim_{t \to \infty} [y(t) - y_*(t)] = 0 \qquad (6.3.8)$$

for any solution $x(t)$ of (6.3.5) with initial conditions $x(0) = x_0 \in \Omega$, where Ω is a given set of initial conditions.

The problem is to find a control function in the form of either an open-loop (feedforward) control

$$u(t) = U(t, x_0) \qquad (6.3.9)$$

or in the form of state feedback

$$u(t) = U(x(t)) \qquad (6.3.10)$$

or output feedback

$$u(t) = U(y(t)) \qquad (6.3.11)$$

to ensure the goal (6.3.7) or (6.3.8).

Such a problem is nothing but a tracking problem standard for control theory. However, the key feature of the control of chaotic systems is to achieve the goal by means of sufficiently small (ideally, arbitrarily small) control. Solvability of this task is not obvious since the trajectory $x_*(t)$ is unstable.

A special case of the above problem is stabilization of the unstable equilibrium x_{*0} of system (6.3.5) with $u = 0$, i.e., stabilization of x_{*0}, satisfying $F(x_{*0}, 0) = 0$. Again, this is just the standard regulation problem with an additional restriction that "small control" solutions are sought. Such a restriction makes the problem far from standard: even for a simple pendulum, nonlocal solutions of the stabilization problem with small control are nontrivial. The class of admissible control laws can be extended by introducing dynamic feedback described by differential or time-delayed models. Similar formulations hold for discrete and time-delayed systems.

Chaotization. A second class of control goals corresponds to the problems of *excitation* or *generation* of chaotic oscillations (also called *chaotization*, *chaotification*, or *anticontrol*). Sometimes these problems can be reduced to the form (6.3.8), but the goal trajectory $x_*(t)$ is no longer periodic, while the initial state is equilibrium. The goal trajectory may be specified only partially. Otherwise, the goal may be to meet some formal criterion of chaos, e.g., positivity of the largest Lyapunov exponent.

Synchronization. Third important class of control goals corresponds to synchronization (more accurately, *controlled synchronization* as opposed to *autosynchronization* or *self-synchronization*). Generally speaking, synchronization is understood as concordance or concurrent change of the states of two or more systems or, perhaps, concurrent change of some quantities related to the systems, e.g., alignment of oscillation frequencies. If the required relation is established only asymptotically, one may speak about *asymptotic synchronization*. If synchronization does not exist in the system without control ($u = 0$) the following *controlled synchronization* problem may be posed: find a control function $u(t)$ ensuring synchronization in the closed-loop system. In this case synchronization is the control goal. For example, the goal corresponding to asymptotic synchronization of the two system states x_1 and x_2 can be expressed as follows:

$$\lim_{t\to\infty}[x_1(t) - x_2(t)] = 0. \tag{6.3.12}$$

In the extended state space $x = \{x_1, x_2\}$ of the overall system, relation (6.3.12) implies convergence of the solution $x(t)$ to the diagonal set $\{x : x_1 = x_2\}$.

Asymptotic identity of the values of some quantity $G(x)$ for two systems can be formulated as

$$\lim_{t\to\infty}[G(x_1(t)) - G(x_2(t))] = 0. \tag{6.3.13}$$

Goal functions. To solve a control problem it is often convenient to rewrite the goals (6.3.7), (6.3.8), (6.3.12), or (6.3.13) in terms of appropriate goal function $Q(x,t)$ as follows:

$$\lim_{t\to\infty} Q(x(t), t) = 0. \tag{6.3.14}$$

For example, to reduce goal (6.3.12) to the form (6.3.14) one may choose the squared Euclidean distance between state vectors of the subsystems as a goal function:

$$Q(x) = |x_1 - x_2|^2.$$

Instead of Euclidean norm other quadratic functions can also be used, e.g., for the case of the goal (6.3.7) the goal function

$$Q(x, t) = [x - x_*(t)]^\mathrm{T} \Gamma [x - x_*(t)],$$

where Γ is a positive definite symmetric matrix can be used. The choice of the matrix Γ provides the possibility of weighting different components of the system state vector to take into account differences in their scale or importance.

In the case of chaotization problem, a goal function $G(x)$ may be introduced such that the goal is to achieve the limit inequality

$$\underline{\lim}_{t\to\infty} G(x(t)) \geq G_*. \tag{6.3.15}$$

Typical choice of the goal function for chaotization is the largest Lyapunov exponent: $G = \lambda_1$ with $G_* > 0$. In some cases the total energy of mechanical or electrical oscillations can serve as $G(x)$.

In terms of goal functions more subtle control goals can be specified, e.g., the control goal may be to modify a chaotic attractor of the free system in the sense of changing some of its characteristics (Lyapunov exponents, entropy, fractal dimension, etc). The freedom of choice of the goal function can be utilized for design purposes.

6.4 Methods of controlling chaos: Continuous-time control

6.4.1 Feedforward control by periodic signal

Methods of *feedforward* control (also called *nonfeedback* or *open-loop* control) change the behavior of a nonlinear system by applying a properly chosen input function $u(t)$ – external excitation. Excitation can reflect influence of some physical action, e.g., external force/field/signal, or it can be some parameter perturbation (modulation). In all cases the value $u(t)$ depends only on time and does not depend on current measurements of the system variables. Such an approach is attractive because of its simplicity: no measurements or extra sensors are needed. It is especially advantageous for ultrafast processes at the molecular or atomic level where no possibility of system variable online measurements exists.

The possibility of significant changes to system dynamics by periodic excitation has been known, perhaps, since the beginning of the 20th century. A. Stephenson discovered in 1908 that a high-frequency excitation can stabilize the unstable equilibrium of a pendulum [417]. Later theoretical results and experiments of P.L. Kapitsa, N.N. Bogoljubov in the 1940s–1950s triggered the development of vibrational mechanics and vibrational control. Analysis of general nonlinear systems affected by high-frequency excitation is based on the Krylov–Bogoljubov averaging method [83]. According to the averaging method stability analysis of a periodically excited system is reduced to analysis of the simplified averaged system. The method provides conditions guaranteeing approximate stabilization of the given equilibrium or the desired (goal) trajectory. A related form of averaging method deals with systems excited by stochastic disturbance (dither). Accuracy of averaging method increases if excitation contains high-frequency harmonics. For physical systems it implies high forcing amplitudes.

In control theory, high-frequency excitation and parameter modulation were studied within the framework of vibrational control [62, 292] and dither control [459]. However, the above-mentioned works dealt only with the problem of stabilizing a given equilibrium or the desired (goal) trajectory. In [305, 308], the use of piecewise constant dither control to modify system dynamics (nonlinearity shape, equilibrium points, etc.) for systems in Lur'e form was proposed. In particular, the creation and elimination of chaotic behavior was studied using heuristic conditions for chaos suggested by [175]. A vast literature is devoted to excitation with medium frequencies – those comparable with the natural frequencies of the system.

The possibility of transforming a periodic motion into chaotic one and vice versa by means of periodic excitation of medium level was demonstrated still in mid-1980s by a group of researchers from Moscow State University [9, 10, 123, 249, 253, 312]. In 1983, K. Matsumoto and I. Tsuda demonstrated the possibility of suppressing chaos in a Belousov–Zhabotinsky reaction by adding a white noise disturbance [288]. Those results were based on computer simulations. In 1988 M.Pettini [347] and in 1990, R. Lima and M. Pettini [270] studied Duffing–Holmes oscillator

$$\ddot{\varphi} - c\varphi + b\varphi^3 = -a\dot{\varphi} + d\cos(\omega t) \qquad (6.4.16)$$

by Melnikov method. The right-hand side of (6.4.16) was considered as a small perturbation of the unperturbed Hamiltonian system. The Melnikov function related to the rate of change of the distance between stable and unstable manifolds for small perturbations was calculated analytically and parameter values producing chaotic behavior of the system were chosen. Then additional excitation was introduced into the parameter of nonlinearity $b \to b(1 + \eta \cos \Omega t)$ and a new Melnikov function was computed and studied numerically. It was shown that if Ω is close to the frequency of initial excitation ω then chaos may be destroyed. Experimental confirmation of this phenomenon was presented in 1991 using a magnetoelastic device with two permanent magnets, an electromagnetic shaker and an optical sensor.

Recent investigations, see e.g. [98, 99, 271, 278] were aimed at better suppression of chaos with smaller values of excitation amplitude and providing convergence of the system trajectories to the desired periodic orbit (limit cycle), better justification of theoretical results. Control of discrete-time systems (maps) and autonomous systems were also studied. Since Melnikov method leads to intractable calculations for state dimensions greater than two, analytical results are known only for periodically excited systems with one degree of freedom. For higher dimensions computer simulations are used. The general problem of finding analytic conditions for creation or suppression of chaos by feedforward periodic excitation of small or medium level still remains open.

In a number of papers the choice of excitation function is based on tailoring it to the system nonlinearity. Let the controlled system be described by equations:

$$\dot{x} = f(x) + Bu, x \in \mathbb{R}^n, u \in \mathbb{R}^m. \qquad (6.4.17)$$

Now let $m = n$ and $\det B \neq 0$. If the desired solution of the controlled system is $x_*(t)$ then an intuitively reasonable choice of excitation is

$$u_*(t) = B^{-1}(\dot{x}_*(t) - f(x_*(t))), \qquad (6.4.18)$$

because $x_*(t)$ will satisfy the equations of the excited system, see [203]. The equation for the error $e = x - x_*(t)$ is then $\dot{e} = f(e + x_*(t)) - f(x_*(t))$. If the linearized system with matrix $A(t) = \partial f(x_*(t))/\partial x$ is uniformly stable in the sense that $A(t) + A(t)^{\mathrm{T}} \leq -\lambda I_n$ for some λ and for all $t \geq 0$ then all solutions of (6.4.17), (6.4.18) will converge to $x_*(t)$ (more general convergence conditions can be found in [164]. In the case when $m < n$ and B is singular the same result is valid under matching conditions: vector $\dot{x}_*(t) - f(x_*(t))$ is in the span of the columns of B. Then the control can be chosen to be $u_*(t) = B^+(\dot{x}_*(t) - f(x_*(t)))$, where B^+ is the pseudoinverse matrix. Despite the fact that the uniform stability condition rules out chaotic (i.e., unstable) trajectories, it is claimed in a number of papers that some local convergence to chaotic trajectories is observed if the instability regions are not dominating.

The applications of feedforward control of chaos to control of CO_2 lasers, Josephson junctions, liquid crystal models, bistable mechanical devices, circular yttrium-ion-garnet films, Murali–Lakshmanan–Chua electronic circuit, FitzHugh–Nagumo equations describing propagation of nerve pulses in a neuronal membrane etc. were reported, see [27, 28].

An interesting application is related to control of the dipole domains. As was demonstrated in [392], propagation of the dipole domains in the GaAs/AlAs-superlattice can be controlled by an external high-frequency field. The doped GaAs/AlAs-superlattice manifests negative conductivity, which leads to propagation of the dipole domains. Various, including chaotic, modes of domain propagation occur depending on the frequency of external action. Frequency locking was shown to be realizable by maintaining the external field frequency within a certain range which extends with the amplitudes of high-frequency voltage. Outside this range, quasiperiodic and chaotic oscillations occur.

Summarizing, a variety of different open-loop methods have been proposed. Most of them were evaluated by simulation for special cases and model examples. However, the general problem of finding conditions for creation or suppression of chaos by feedforward excitation still remains open.

6.4.2 Linearization of Poincaré map (OGY method)

A real burst of interest in the control of chaotic systems was caused by the paper by E. Ott, C. Grebogi, and J. Yorke [331] published in 1990. In this paper two key ideas were introduced:

1. To use the discrete system model based on linearization of the Poincaré map for controller design.
2. To use the recurrence property of chaotic motions and apply control action only at time instants when the motion returns to the neighborhood of the desired state or orbit.

Numerous extensions and interpretations have been proposed by different authors in subsequent years and the method is commonly referred to as the "OGY method." The essence of the OGY method is as follows.

Let the controlled system be described by the state space equations

$$\dot{x} = F(x, u), \qquad (6.4.19)$$

where $x \in \mathbb{R}^n, u \in \mathbb{R}^1$. Let the desired (goal) trajectory $x_*(t)$ be a solution of (6.4.19) with $u = 0$. The goal trajectory may be either periodic or chaotic: in both cases it is recurrent. Draw a surface (Poincaré section)

$$S = \{x : s(x) = 0\} \qquad (6.4.20)$$

through the given point $x_0 = x_*(0)$ transversely to the solution $x_*(t)$ (i.e., $\partial s(x)/\partial x \, F(x, 0) \neq 0$). Consider the map $x \mapsto P(x, u)$ where $P(x, u)$ is the point of first return to S of the solution to (6.4.19) with constant input u started from x. The map $x \mapsto P(x, u)$ is called *the controlled Poincaré map*. It is well defined at least in some vicinity of the point x_0 owing to the recurrence property of $x_*(t)$. Iterating the map leads to a discrete-time system

6.4 Methods of controlling chaos: Continuous-time control

$$x_{k+1} = P(x_k, u_k), \qquad (6.4.21)$$

where $x_k = x(t_k)$, t_k is the time of the kth crossing and u_k is the value of $u(t)$ between t_k and t_{k+1}. Another way of creating a discrete-time model of controlled system is applying a pulse control action rather than a piecewise constant one: $u(t) = u_k/\varepsilon$ for $t_k \leq t \leq t_k + \varepsilon$, $u(t) = 0$ for $t_k + \varepsilon < t < t_{k+1}$. For small $\varepsilon > 0$ it leads to an affine in control model $x_{k+1} = f(x_k) + g(x_k)u_k$.

The next step of the control law design is to replace the initial system (6.4.19) by the linearized discrete system

$$\tilde{x}_{k+1} = A\tilde{x}_k + Bu_k, \qquad (6.4.22)$$

where $\tilde{x}_k = x_k - x_0$ and to find a linear stabilizing controller $u_k = C\tilde{x}_k$ for (6.4.21) by means of one of standard methods of linear control theory. Finally, the control law is endowed with the switching off rule applied when the error exceeds some threshold value Δ as follows:

$$u_k = \begin{cases} C\tilde{x}_k, & \text{if } |\tilde{x}_k| \leq \Delta, \\ 0, & \text{otherwise.} \end{cases} \qquad (6.4.23)$$

A key point of the method is to apply control only in some vicinity of the goal trajectory by introducing the switching off threshold ("outer" deadzone). This has the effect of bounding control action and allows to respect small control restriction, based on the special recurrence feature of chaotic motions.

Efficiency of such an approach has been confirmed by numerous simulations performed by different authors. However, the convergence rate may be low that is the price of achieving nonlocal stabilization of a nonlinear system by small control.

There are two important problems making implementation of the method difficult: lack of information about the system model and incomplete measurements of the system state. The second difficulty can be overcome by replacing the initial state vector x by the so-called *delayed coordinates vector* $X(t) = [y(t), y(t-\tau), \ldots, y(t-(N-1)\tau)]^T \in \mathbb{R}^n$, where $y = h(x)$ is the output (e.g. one of the system coordinates) available for measurement and $\tau > 0$ is delay time. Then the control law has the form:

$$u_k = \begin{cases} \mathcal{U}(y_k, y_{k,1}, \ldots, y_{k,N-1}), \\ \quad \text{if } |y_{k,i} - y_*| \leq \Delta, \ i = 1, \ldots, N-1, \\ 0, \quad \text{otherwise,} \end{cases} \qquad (6.4.24)$$

where $y_{k,i} = y(t_k - i\tau)$, Δ is the switching off threshold, $\mathcal{U}(y_k, y_{k,1}, \ldots, y_{k,N-1})$ is the control law designed using one of the methods of linear control theory.

A special case of algorithm (6.4.24) introduced by E. Hunt [204] in 1991 was termed *occasional proportional feedback* (OPF). The OPF algorithm is used for stabilization of the amplitude of a limit cycle and is based on measuring local maxima (or minima) of the output $y(t)$, i.e., the Poincaré section S is defined as in (6.4.20) with $s(x) = \partial h/\partial x\, F(x,0)$. Obviously, crossing of S occurs when $\dot{y}(t) = 0$. If y_k is the value of the kth local maximum, then the OPF method suggests a simple control law

$$u_k = \begin{cases} K\tilde{y}_k, & \text{if } |\tilde{y}_k| \leq \Delta, \\ 0, & \text{otherwise,} \end{cases} \quad (6.4.25)$$

where $\tilde{y}_k = y_k - y_*$ and $y_* = h(x_0)$ is the desired upper level of oscillations.

To overcome the first problem – uncertainty of the linearized plant model it was suggested by Ott–Grebogi–Yorke and their followers to estimate parameters of the Poincaré map (in state-space form or in the form depending on the delayed coordinates).

The problem of parameter estimation (identification) falls within the scope of the identification theory. It is well known that using identification algorithms in closed-loop (i.e., together with control algorithms) may prevent from "good" estimation. Therefore, justification of the algorithms (6.4.24) and (6.4.25) is hard.

A new version of identification algorithm allowing for its justification was proposed by A. Fradkov and P. Guzenko in 1997 [150, 152, 164] for the special case when $y_{k,i} = y_{k-i}$, $i = 1, \ldots, n$. In this case the outputs are measured and control action is changed only when the trajectory crosses the surface. The input–output model has the following form:

$$y_k + a_1 y_{k,1} + \cdots + a_{N-1} y_{k,N-1} = b_1 u_k + \cdots + b_{N-1} u_{k-N-1}. \quad (6.4.26)$$

The model (6.4.26) contains fewer coefficients than the state-space model (6.4.22). It facilitates the controller design. For estimation, the method of recursive goal inequalities due to V.A. Yakubovich was used, allowing to resolve the problem of identification in the closed-loop by means of introducing an additional inner deadzone. An inner deadzone combined with outer deadzone of the OGY method provides robustness of the identification-based control with respect to both model errors and measurements errors. The Fradkov–Guzenko algorithm suggests the standard model reference based linear control law

$$\bar{u}_k = b_1^{-1}\big((a_1-g_1)y_{k,1}+\cdots+(a_N-g_N)y_{k,N-1}-b_2 u_{k-1}-\cdots-b_{N-1}u_{k-N-1}\big), \quad (6.4.27)$$

where g_i, $i = 1, 2, \ldots, N-1$ are coefficients of the *reference model*, chosen to ensure stability of the polynomial $G(\lambda) = \lambda^{N-1} + g_1 \lambda^{N-2} + \cdots + g_{N-1}$. In a more compact vector form the law (6.4.27) is as follows:

$$\bar{u}_k = \theta_k^\mathrm{T} X(t_k), \quad (6.4.28)$$

where $\theta_k \in \mathbb{R}^{2(N-1)}$ is the vector of the current estimates of the controller (6.4.27) parameters, $X(t_k)$ is the vector of the delayed coordinates. Parameter estimates are adjusted according to the adaptation algorithm

6.4 Methods of controlling chaos: Continuous-time control

$$\mu_{k+1} = \begin{cases} 1, & \text{if } |y_{k+1} - y_*| > \Delta_y \text{ and} \\ & |y_{k-i} - \overline{y}(t_{k-i})| < \overline{\Delta}, \quad i = 0, \ldots, N-1, \\ 0, & \text{else}; \end{cases}$$

$$\theta'_{k+1} = \begin{cases} \theta_k - \gamma(\text{sign} b_0)(y_{k+1} - y_*)w_k/\|w_k\|^2, \\ \qquad \text{if } \mu_{k+1} = 1, \\ \theta_k \qquad \text{else}; \end{cases}$$

$$u'_{k+1} = \theta'^{\mathrm{T}}_{k+1} w_{k+1}$$

$$\theta_{k+1} = \begin{cases} \theta'_{k+1} & \text{if } |u'_{k+1}| \leq \overline{u} \text{ and } \mu_{k+1} = 1, \\ \theta'_{k+1} - (u'_{k+1} - \overline{u})/\|w_k\|^2, & \text{if } u'_{k+1} > \overline{u} \text{ and } \mu_{k+1} = 1, \\ \theta'_{k+1} - (u'_{k+1} + \overline{u})/\|w_k\|^2, & \text{if } u'_{k+1} < -\overline{u} \text{ and } \mu_{k+1} = 1, \\ \theta_k, & \text{if } \mu_{k+1} = 0. \end{cases}$$

(6.4.29)

where $\gamma > 0$ is the adaptation gain, $\overline{u}$ is maximum admitted value of control; Δ_y is maximum admitted value of the difference between y_k and y_*. The value of the threshold $\overline{\Delta}$ is related to the size of the "tube" around the base trajectory $\overline{x}(t)$, where the input–output model (6.4.26) is defined.

The above adaptive control algorithm is justified for recurrent systems under the so-called n-observability condition, ensuring that the state remains small is the output of the controlled system is small at some n consecutive crossings.

Further modifications and extensions to the OGY method have been recently proposed extending the basin of attraction and reducing the transient time: a multi-step version; a quasicontinuous extension; iterative refinement, etc.

Efficiency of the OGY method has been demonstrated by physical experiments with magnetoelastic ribbon [120], glow discharge, nonautonomous RL-diode circuit. The OPF method has been used for stabilization of the frequency emission from a tunable lead-salt stripe geometry infrared diode laser and implemented in an electronic chaos controller, see references in [27].

6.4.3 Delayed feedback (Pyragas method)

The method of time-delayed feedback suggests to find and stabilize a τ-periodic orbit of the nonlinear system (6.3.5) by a simple control action

$$u(t) = K[x(t) - x(t - \tau)] \qquad (6.4.30)$$

where K is feedback gain, and τ is time-delay. If τ is equal to the period of an existing periodic solution $\overline{x}(t)$ of (6.3.5) for $u = 0$ and the solution $x(t)$ to the closed-loop system (6.3.5), (6.4.30) starts from $\Gamma = \{\overline{x}(t)\}$, then it will remain in Γ for all $t \geq 0$. A puzzling observation was made however, that $x(t)$ may converge to Γ even if $x(0) \bar{\in} \Gamma$.

The law (6.4.30) applies also to stabilization of forced periodic motions in system (6.3.5) with a T-periodic right-hand side. Then τ should be chosen equal to T. The formulation of the method for stabilization of fixed points and periodic solutions of discrete-time systems is straightforward. The method was proposed by Lithuanian physicist K. Pyragas in 1992 and is also called the Pyragas method [362].

An extended version of the Pyragas method has also been proposed with

$$u(t) = K \sum_{k=0}^{M} r_k [y(t - k\tau) - y(t - (k+1)\tau] \qquad (6.4.31)$$

where $y(t) = h(x(t)) \in \mathbb{R}^1$ is the observed output and $r_k, k = 1, \ldots, M$ are tuning parameters. For $r_k = r^k, |r| < 1$, and $M \to \infty$ the control law (6.4.31) tends to

$$u(t) = K[y(t) - y(t - \tau)] + Kru(t - \tau) \qquad (6.4.32)$$

Despite simplicity of the algorithms (6.4.30)–(6.4.32), analytical study of the closed-loop behavior is difficult. T. Ushio [431] established a simple necessary condition ("oddness limitation") for stabilizability with a Pyragas controller (6.4.30) for a class of discrete-time systems. Proofs for more general and continuous-time cases were given independently in [217, 310]. The result is as follows. Let $\Phi(t)$ be the fundamental matrix of the system (6.3.5) linearized along given τ-periodic solution $x_*(t)$. Evaluate the matrix $\Phi(\tau)$ (called *monodromy matrix* or *Floquet matrix*) and then the eigenvalues of $\Phi(\tau)$ (called *multipliers*) μ_i, $i = 1, 2, \ldots, n$ that are linked to the Lyapunov exponents ϱ_i of the τ-periodic solution $x_*(t)$ by the relations $\varrho_i = \tau^{-1} \ln |\mu_i|$.

The above-mentioned necessary condition is that the number of real unstable multipliers (eigenvalues of the matrix $\Phi(\tau)$ with absolute value exceeding 1) should not be odd.

Just et al. [218] gave a more detailed analysis and established approximate bounds for a stabilizing gain K. It is worth to notice that the set of the values of K providing stablization [217, 218] includes small values of K for small instablity degree $\max \varrho_i$, and vanishes for sufficiently large instablity degree $\max \varrho_i$ of the linearized system. H. Schuster and M. Stemmler [393] showed that the "oddness limitation" can be relaxed by means of a periodic modulation of the gain K.

More information about dynamics of the control system is available for a generalized Pyragas controller, proposed by M. Basso, R. Genesio, and A. Tesi [55, 56]:

$$u(t) = G(p)[y(t) - y(t - \tau)] \qquad (6.4.33)$$

where $G(p)$ is transfer function of the linear time-invariant filter, $p = d/dt$. System (6.3.5), (6.4.33) was considered as a *Lurie system* (system represented as feedback connection of a linear dynamical part and a (static or dynamic) nonlinearity) and studied by the methods of absolute stability theory [265, 266, 451]. It allowed to obtain sufficient conditions on the transfer

6.4 Methods of controlling chaos: Continuous-time control

function of the linear part and on the slope of nonlinearity under which there exists stabilizing law (6.4.33). In [56], the procedures for "optimal" controller design, maximizing the stability bound as well as extension to systems with a nonlinear nominal part were proposed.

Choosing $|r| > 1$, in (6.4.32) one obtains the algorithm with an unstable internal dynamics. However, such an algorithm is realizable if the closed-loop system is stable. It was shown by K. Pyragas [363] that using an unstable controller allows to significantly weaken limitations for the monodromy matrix $\Phi(t)$ and, particularly, to remove the "oddness" limitation.

The Pyragas method was extended to coupled (open flow) systems, modified for systems with symmetries, and extended to include an observer estimating the difference between the system state and the desired unstable trajectory (fixed point).

Reported applications include stabilization of coherent modes of lasers, magnetoelastic systems, control of cardiac conduction model, control of stick-slip friction oscillations, traffic models, PWM controlled buck converter, paced excitable oscillator described by the FitzHugh–Nagumo model, catalytic reactions in bubbling gas–solid fluidized bed reactors.

A drawback of the control law (6.4.30) is its sensitivity to parameter choice, especially to the choice of the delay τ. Apparently, if the system is T-periodic and the goal is to stabilize its forced T-periodic solution, then the choice $\tau = T$ is mandatory. If the period T is unknown, a heuristic trick can be used: to simulate the unforced system with initial condition $x(0)$ until the current state $x(t)$ approaches $x(s)$ for some $s < t$, i.e., until $|x(t) - x(s)| < \varepsilon$ with a sufficiently small ε. Then the choice $\tau = t - s$ will give a reasonable estimate of a period and the vector $x(t)$ will be an initial condition to start control. However, such an approach often gives overly large values of the period.

An alternative is to introduce the adaptation of the delay time τ. A. Kittel, J. Parisi, and K. Pyragas proposed in [235] the following adaptive delayed feedback for the case when the Poincaré section S transverse to the desired trajectory is additionally given: $u(t) = K[y(t) - y(t - \tau_k)]$, where $\tau_k = t_k - t_{k-d}$ is changing at the time t_k of the kth crossing of S (e.g., time of the kth maximum of the output $y(t)$). Specifying an integer d one may stabilize a d-periodic orbit of the Poincaré map.

Since chaotic attractors contain periodic solutions of different periods, an important problem is to find and to stabilize (with small control) the solution with the smallest period. This problem remains open.

6.4.4 Linear and nonlinear control

Many standard control engineering notions and techniques are suitable for the control of chaos. In some cases even the simple proportional feedback,

$$u(t) = KB^+(x - x_*(t)), \qquad (6.4.34)$$

or more sophisticated *open-plus-closed-loop* (OPCL) algorithm [212]

$$u(t) = B^+[\dot{x}_*(t) - f(x_*(t)) - K(x - x_*(t))]. \tag{6.4.35}$$

where K is gain matrix, $x_*(t)$ is the desired (reference) trajectory, B^+ is pseudoinverse matrix, allow to achieve the desired control goal for controlled systems $\dot{x}(t) = f(x(t)) + Bu$. If the goal trajectory has bounded velocity, the necessary condition for its stabilization is the stability of the matrix $A(t) + BKB^+$, where $A(t) = \frac{\partial F}{\partial x}(\bar{x}_*(t))$ is the matrix of the system model linearized near the desired trajectory. If the matrix $A(t)$ is unstable (which is common for chaotic systems), then stability of $A(t) + BKB^+$ can be easily ensured for $\dim x = \dim u$.

However, in most practical applications the number of controls is less than the number of system states. Then numerous methods of modern nonlinear control can be applied. Below two large classes of nonlinear control methods will be presented: feedback linearization and goal-oriented techniques.

Feedback linearization. To perform feedback linearization means to find a smooth coordinate change $z = \Phi(x)$, $x \in \Omega$ and a feedback transformation of the control variable $u = \alpha(x) + \beta(x)v$, such that in new variables the controlled system dynamics is described by a linear state space model $\dot{z} = Az + Bv$. The theory of feedback linearization was briefly presented in Section 2.4.3. Below we illustrate it by example: control of the Lorenz system.

Example 6.1. (Feedback linearization control of the Lorenz system). Consider the controlled Lorenz system with control appearing in the third equation:

$$\dot{x}_1 = \sigma(x_2 - x_1),$$
$$\dot{x}_2 = rx_1 - x_2 - x_1 x_3,$$
$$\dot{x}_3 = -\beta x_3 + x_1 x_2 + u.$$

Let $y = x_1$. Then

$$L_f y = \dot{y} = \dot{x}_1 = \sigma(x_2 - x_1),$$
$$L_f^2 y = L_f(L_f y) = \ddot{x}_1 = \sigma(\dot{x}_2 - \dot{x}_1) = \sigma\big[(r+1)x_1 - 2x_2 + x_1 x_3\big],$$

and, therefore, relative degree is equal to 3 everywhere except the plane $x_1 = 0$. New coordinates can be chosen as follows:

$$z = \Phi(x): \quad z_1 = x_1,$$
$$z_2 = \sigma(x_2 - x_1),$$
$$z_3 = \sigma\big[(r+1)x_1 + 2x_2 + x_1 x_3\big],$$
$$x = \Phi^{-1}(z): \quad x_1 = z_1,$$
$$x_2 = \frac{1}{\sigma} z_2 + z_1,$$
$$x_3 = \frac{1}{x_1}\left[\frac{1}{\sigma} z_3 - (r-1)z_1 - \frac{2}{\sigma} z_2\right].$$

6.4 Methods of controlling chaos: Continuous-time control

It is seen that the system is feedback linearizable for $x_1 \neq 0$. Thus for Lorenz system there is no globally defined smooth feedback linearizing transformation. Feedback linearization allows to stabilize the system to any fixed point in any half-space $\{x_1 < 0\}$, $\{x_1 > 0\}$ is not suitable for global stabilization of the Lorenz system.

Feedback linearization was applied to control of chaotic systems, e.g., in [14, 49, 105, 456]. Its drawback is in that the approach ignores the internal dynamics of the system and formally allows to achieve any desired dynamics of the closed-loop system. In fact, the achievement of arbitrary dynamical behavior may require significant power of control, e.g., if the initial state is far from the desired one or the desired motion is rapidly changing. Such a drawback is typical for a number of works based on conventional control theory approaches.

Another problem is that of incomplete measurements. Standard approach to output feedback control is using an observer-based controller that allows for systematic use of dynamic output feedback. Proportional feedback in the extended space (x, u) (i.e., dynamic feedback) aimed at achievement of the desired dynamics of the closed-loop system was proposed and examined in [282, 461].

The potential of dynamic feedback can be better exploited using an observer-based framework that allows for systematic use of output feedback. A survey of nonlinear observer techniques can be found in [315] (see also Section 5.2). Linear high-gain observer-based control for globally Lipschitz nonlinearities was studied in [268].

Note that models of chaotic systems often do not satisfy a global Lipschitz condition owing to the presence of polynomial nonlinearities $x_1 x_2$, x^2, etc. Although trajectories of chaotic systems are bounded, this is not necessarily the case when the system is influenced by control. Therefore, special attention should be paid to providing boundedness of the solutions by appropriate choice of controls. Otherwise the solution may escape in finite time and it does not make sense to discuss stability and convergence issues. The possibility of escape in nonlinear controlled systems is often overlooked in application papers.

Goal-oriented techniques. A number of methods are based on reduction of the current value of some goal (objective) function $Q(x(t), t)$. The current value $Q(x(t), t)$ may reflect the distance between the current state $x(t)$ and the current point of the goal trajectory $x_*(t)$, such as $Q(x, t) = |x - x_*(t)|^2$, or the distance between the current state and the goal surface $h(x) = 0$, such as $Q(x) = |h(x)|^2$. For continuous-time systems the value $Q(x)$ does not depend directly on control u and decreasing the value of the speed $\dot{Q}(x) = \partial Q/\partial x F(x, u)$ can be posed as immediate control goal instead of decreasing $Q(x)$. This is the basic idea of the *speed-gradient* (SG) method, see Section 2.4.2 that was first used for control of chaotic systems in [137, 162, 163].

Example 6.2. (Stabilization of the equilibrium point of the thermal convection loop model). One of the simplest experimental setup which can demonstrate complex oscillatory behavior is the chaotic thermal convection loop. In the literature the following controlled thermal convection loop model was considered [411]:

$$\begin{aligned} \dot{x} &= \sigma(y - x), \\ \dot{y} &= -y - xz, \\ \dot{z} &= -z + xy - r + u, \end{aligned} \quad (6.4.36)$$

where u is the control variable which is a fluctuation in the heating rate superimposed on the nominal rate r, σ is the Prandtl number and r is the Raleigh number. This model can be obtained from the Lorenz system [276] by replacing $z - r$ with z and assuming that $r = \text{const}$ and $b = 1$. For $u = 0$ and $0 < r < 1$ the system has one stable globally attracting equilibrium $(0, 0, -r)$ that corresponds to the no-motion state of the thermal convection. At $r = 1$, two additional equilibrium points C_+ and C_- emerge: $x = y = \pm\sqrt{r-1}, z = -1$. The convection equilibria lose their stability in the Andronov–Hopf bifurcation at $r = \sigma(\sigma + 4)/(\sigma - 2)$. For larger values of the parameter r the system has no more equilibrium points.

In [411], the on–off controller was proposed to stabilize the inherent unstable equilibrium point of this system:

$$u = -\gamma \text{sgn}(z + 1). \quad (6.4.37)$$

Practical experimentation showed that the controller (6.4.37) stabilizes the thermal convection in either clockwise or counterclockwise direction that corresponds to the stabilization of one of the equilibria C_+ or C_-.

It is easy to see that the controller (6.4.37) is a special case of the speed-gradient algorithm in finite form (2.4.51) for the objective function

$$Q(x, y, z) = (x - \sqrt{r-1})^2/\sigma + (y - \sqrt{r-1})^2 + (z + 1)^2.$$

Indeed, calculating the time derivative of Q along trajectories of (6.4.36) yields

$$\dot{Q}(x, y, z, u) = \omega(x, y, z, u) = -(x - y)^2 + \left(\sqrt{r-1}(x - \sqrt{r-1}) + u\right)(z + 1). \quad (6.4.38)$$

Evaluating partial derivative of (6.4.38) with respect to u and choosing $\psi(z) = \varepsilon \text{sgn}(z + 1)$ one can notice that the control law (6.4.37) is a particular case of the relay algorithm (2.4.51). It was shown in [164] that any trajectory of the overall system tends to some rest point contained in the set of points (x, y, z) such that

$$\left\{ x = y, \ \left|(x + \sqrt{r-1})(x - \sqrt{r-1})\right| \leq \gamma, \ z = -1 \right\}. \quad (6.4.39)$$

It yields convergence of the solution to the vicinity of one of the inherent equilibrium points C_+ or C_-. The size of the limit vicinity is of order γ. Hence for small γ the limit point can be put arbitrarily close to the desired equilibrium. However, the price of good accuracy is long transition time which is inversely proportional to γ for small γ.

Other methods. For stabilization of a goal point or manifold other methods of modern nonlinear control theory have been used, e.g., center manifold theory backstepping iterative design or the method of macrovariables; passivity-based design; absolute stability theory; H_∞ control; combination of Lyapunov and feedback linearization methods, see surveys [27, 28, 153].

A number of papers is devoted to application of variable structure systems (VSS) and sliding modes [239, 453, 457]; Note that VSS algorithms for the switching surface $h(x) = 0$ coincide with the speed-gradient algorithms for the goal function $Q(x) = |h(x)|$.

A fruitful direction is the use of frequency-domain methods applied to nonlinear control. In particular, approximate methods of harmonic balance for evaluation and prediction of chaotic modes can be used together with rigorous absolute stability theory [57]. An interesting method within this framework employs a selective ("washout") filter which damps all signals with frequencies beyond some narrow range [295]. If such a filter is included in the feedback loop of a chaotic system and the base frequency of the filter coincides with the frequency of one of the existing unstable periodic solutions, then it is plausible that the system will be in a periodic motion rather than in chaotic. This approach was applied to control of lasers.

The majority of nonlinear control approaches can be grouped into two large classes: Lyapunov approaches (speed-gradient, passivity-based methods) and compensation approaches (feedback linearization, geometric methods, etc.). The interrelation between these classes can be illustrated as follows. Let the control goal be stabilization of some output variable $y = h(x)$ of the affine system $\dot{x} = f(x) + g(x)u$, at zero level. Lyapunov (or speed-gradient) methods introduce a goal function $Q(x) = |h(x)|^2$ and gradually decrease its derivative $\dot{Q}$ according to the condition $h^T \partial h / \partial x (f + gu) < 0$, e.g., moving along the speed-gradient (antigradient of $\dot{Q}$):

$$u = -\gamma g^T (\nabla h) h.$$

To respect the "small control" requirement it is necessary to choose sufficiently small gain $\gamma > 0$.

On the other hand, the compensation approaches introduce an auxiliary macrovariable $\alpha(x) = \dot{y} + \varrho y$ with some $\varrho > 0$ and immediately force it to zero with the control:

$$u = -\frac{f^T(\nabla h) + \varrho h}{g^T(\nabla h)}.$$

Note that $\alpha = 0$ if and only if $\dot{Q} = -2\varrho Q$, i.e., compensation is equivalent to specifying a rate decrease of $Q(x)$. As a result, any desired "instantaneous" transient rate can be achieved at the cost of loss of flexibility and loss of the "small control" property.

Therefore using the well-developed machinery of modern linear and nonlinear control theories often does not take full account of the special aspects of chaotic motions. This often means that the "small control" requirement is violated. To respect the "small control" requirement the gain $\gamma > 0$ should be sufficiently small. An *outer* deadzone may be introduced in terms of the goal function, e.g.,

$$u(t) = \begin{cases} -\gamma \nabla_u \dot{Q}(x,u), & \text{if } |Q(x(t))| \leq \Delta, \\ 0, & \text{otherwise.} \end{cases} \quad (6.4.40)$$

Another peculiarity of chaotic systems is that the models of chaotic systems often do not satisfy global Lipschitz condition owing to the presence of polynomial nonlinearities $x_1 x_2$, x^2, etc. Although trajectories of chaotic systems are bounded, it is not necessarily the case when the system is influenced by control. Therefore a special attention should be paid to providing boundedness of the solutions by special choice of controls. Otherwise the solution may escape in finite time and it will not make sense to discuss stability and convergence issues.

6.4.5 Adaptive control

In a variety of physical applications parameters of the system under control are unknown. Information about the structure of the model may also be incomplete. It makes adaptive control schemes very promising. Most methods belong to either direct or indirect (identification based) parametric adaptive control schemes. It means that the model of the system is represented in a parametric form:

$$\dot{x} = F(x, \theta, u), y = h(x), \quad (6.4.41)$$

where θ is a vector of unknown parameters. Based on (6.4.41), a parametric representation of the controller is

$$u = \mathcal{U}(x, u, \xi), \quad (6.4.42)$$

where ξ depends on θ, i.e., $\xi = \Phi(\theta)$ for some mapping $\Phi(\cdot)$.

Measured time series of states $\{x(t)\}$ or outputs $\{y(t)\}$ are utilized (offline or online) to evaluate adaptation (tuning) parameters which are estimates of either the system parameters $\hat{\theta}(t)$, or the controller parameters $\hat{\xi}(t)$.

Two approaches can be used for choosing the adaptation parameters: *direct* and *indirect (identification based)* ones. In the direct approach the adaptation parameters are the parameters θ in (6.4.41); in the identification approach the vector of adjustable parameters is the estimates $\hat{\xi}$ of the unknown parameters ξ. In the identification approach the equation of the tunable model is often used:

$$\dot{x}_m(t) = F(x_m, u, t, \hat{\xi}). \quad (6.4.43)$$

For design of adaptation algorithms an auxiliary control objective which expresses the desired dynamics of the plant (6.4.41) and tunable model (6.4.43) can be used. For example, to formalize the goal "variable x_m must tend to x," the goal functional

$$Q(x_m, x, t) = Q(x - x_m(t)), \qquad (6.4.44)$$

can be used which includes explicitly the solution $x_m(t)$ of (6.4.43) (the case of *explicit tunable model*). Instead of (6.4.44), it is also possible to choose the goal function

$$\tilde{Q}(x, \hat{\xi}, t) = \tilde{Q}(F(x(t), u(t), t, \xi) - F(x_m(t), u(t), t, \hat{\xi})) \qquad (6.4.45)$$

(the case of *implicit tunable model*). In both cases (direct and indirect schemes) $Q(x, \hat{\xi}, t)$ is a nonnegative smooth scalar function. In both cases one can write out the so-called *generalized plant equation*:

$$\dot{\tilde{x}} = \tilde{F}(\tilde{x}, t, \theta) \quad \text{or} \quad \dot{\tilde{x}} = \tilde{F}(\tilde{x}, t, \hat{\xi}), \qquad (6.4.46)$$

which can be obtained by substitution of (6.4.42) into (6.4.41) and assuming that $\varphi \equiv 0$. In the direct approach the adaptation algorithm is given by

$$\theta(t) = \Theta'[x(s), u(s), \theta(s), 0 \leq s \leq t]. \qquad (6.4.47)$$

In the identification approach the parameter update law has the following form:

$$\hat{\xi}(t) = \Theta''[x(s), u(s), \hat{\xi}(s), 0 \leq s \leq t]. \qquad (6.4.48)$$

It is worth mentioning that sometimes it is difficult to distinguish direct and identification approaches. Moreover, the designer may combine direct and identification approaches in order to improve the system performance.

For the adaptation (tuning) of parameters a variety of existing methods can be used such as gradient, speed-gradient, least-squares, weighted least-squares, etc. For the continuous-time case a wide class of adaptation algorithms are encompassed by speed-gradient algorithm in differential form (2.4.40). The input data for the design procedure are the generalized plant equation (6.4.46) and the control objective (in the case of the identification approach one has to take the new objective (6.4.44) or (6.4.45)). If the influence of the disturbances cannot be neglected, robustified versions of the algorithms may be used.

Most existing results are based on linearly parameterized models (6.4.41) or linearly parameterized controllers (6.4.42). The controller (6.4.42) is usually designed using model reference or feedback linearization approaches. Proofs are typically based on Lyapunov functions, quadratic in original or in some transformed variables.

6.5 Discrete-time control

Some discrete-time algorithms were already mentioned in Section 6.4.2 (when discussing methods based on the Poincaré map) and in Section 6.4.3. They can be considered as some kind of sample-data control. There are many results on stability of sample-data feedback control systems in the literature that may be employed for chaos control.

Although many authors use the term "optimal control," in most cases only *locally optimal* solutions are proposed, based on minimization over u of one-step-ahead losses $Q(F_d(x_k,u),u)$, where F_d comes from plant model (2.1.11) and $Q(x,u)$ is a cost function, e.g., $Q(x,u) = |x-x_*|^2 + \kappa|u|^2$, see [1]. The choice of a large weight $\kappa > 0$ allows enforcement of the "small control" requirement. For large κ locally optimal control is close to the gradient $u_{k+1} = -\gamma \nabla_u Q(F_d(x_k,u),u)$, with small $\gamma > 0$ [164].

A substantial number of the papers devoted to discrete-time control of chaos deal with low-order examples. The variety of discrete-time examples of chaotic systems seems even broader than that of continuous-time ones owing to a number of one- and two-dimensional systems that do not have continuous-time counterparts (this follows from Poincaré–Bendixon theorem stating that a smooth autonomous differential system evolving on a two-dimensional manifold may have only equilibria or limit cycles as ω-limit sets, i.e., cannot be chaotic). Among popular examples for discrete-time control of chaos are systems described by the logistic map [110, 128, 427]:

$$x_{k+1} = ax_k(1-x_k);$$

the Hénon system [191]

$$x_{k+1} = 1 - ax_k^2 + y_k, y_{k+1} = -Jx_k;$$

the tent map [351]

$$x_{k+1} = rx_k, 0 \leq x_k < 0.5; x_{k+1} = r(1-x_k), 0.5 \leq x_k \leq 1$$

the standard (Chirikov) map [251]

$$v_{k+1} = v_k + K sin\phi_k, \phi_{k+1} = \phi_k + v_k.$$

Only a few results are available for multidimensional systems. They are based upon the gradient method [1, 164], variable structure systems [269], generalized predictive control [335].

6.6 Generation of chaos (chaotization)

The problem of chaotization of a given system by feedback (called also chaos synthesis, chaos generation, anticontrol of chaos or chaotification) appears when it is necessary to design (generate) chaotic signals, e.g., for information encryption, broadband communications, computation with pseudorandom numbers (Monte-Carlo method), etc. Studying ways of creating chaotic signals with prescribed properties may shed light upon mechanisms of biological systems, e.g., cardiac, and brain activities. The problem was first considered by A. Vanecek and S. Celikovsky [434] in 1994, who proposed scenario of chaotization for Lurie systems represented by a transfer function with poles $s_1, \ldots, s_n$, zeros $z_1, \ldots, z_{n-1}$ and an odd monotone nonlinearity in a feedback circuit. Vanecek–Celikovsky scenario suggests to choose poles and zeros in order to provide the characteristic polynomial

$$(\lambda - s_1) \cdots (\lambda - s_n) + k(\lambda - z_1) \cdots (\lambda - z_{n-1})$$

of the linear part closed by a linear feedback $u = ky$ for all $k : 0 < k < \infty$, with the following properties: partial instability (roots with both negative and positive real parts are present); hyperbolicity (no roots have zero real part); dissipation (sum of the poles is negative); nonpotentiality (some poles have nonzero imaginary parts). Chaos under such conditions is related to the Shilnikov theorem [189, 402] (Shilnikov chaos).

A different approach was proposed by G. Chen and D. Lai [104] in 1998 and then by X. Wang and G. Chen [444] in 1999. It is based on the Marotto theorem, providing generalization of Sharkovsky–Li–Yorke criterion of chaos to multidimensional discrete systems (see Section 6.2). For discrete systems of the form $x_{k+1} = f(x_k) + u_k$ a feedback $u_k = \varepsilon g(\sigma x_k)$ is sought, where $\varepsilon > 0, \sigma > 0$, and the graph of $g(x)$ has a saw-toothed or sinusoidal shape. Let $x = 0$ be the stable equilibrium of the free system ($|f'(x)| < 1$ for $x \in \Delta = (-\varrho, \varrho)$ for some $\varrho > 0$). Then choosing sufficiently large σ one can make the equilibrium $x = 0$ of the closed-loop system $x_{k+1} = f_c(x_k)$, where $f_c(x) = f(x) + \varepsilon g(\sigma x)$, unstable ($|f'_c(x)| > 1$ for $x \in \Delta$), at the same time providing convergence to zero of at least one trajectory in finite time: $f_c^m(x_0) = 0$ for some $x_0 \in \Delta$ and some integer $m > 0$ with $(f^m(x_0))' \neq 0$. Chaoticity follows from the Marotto theorem. It is important that the maximum value of $|u_k|$ can be made less than any positive constant by means of proper choice of ε. This method was applied, e.g., to design of chaotic neural networks.

6.7 Time and energy needed for control of chaos

An important open problem is to evaluate time and energy required for control. Despite the common belief that small control allows the suppression of strong chaos, only a few analytic results are available. The first result of such kind was obtained in the pioneer paper [331], where the time required for stabilizing an unstable trajectory with low intensity control was evaluated for second-order systems. It was shown that required time depends on control intensity according to power law. Later a new control algorithm based on symbolic dynamics was proposed [229] ensuring a logarithmic law for hyperbolic systems and power law for neutral systems [230, 231].

Below some estimates of time and energy required for control of chaotic motions are outlined, including estimates of transient time and traveling time over attractor. It is shown that existing results provide either power or logarithmic dependence of transient time on control intensity.

Consider a dynamical system described by differential equation

$$\dot{x} = f(x, u), \tag{6.7.49}$$

where x is n-dimensional vector of the state variables; Let $x(t, u, x_0)$ be the solution of the system (6.7.49) with initial condition $x(0) = x_0$. Let the system (6.7.49) for $u(t) \equiv 0$ possess a chaotic trajectory $\tilde{x}(t)$, $\tilde{x}(0) = \tilde{x}_0$. A typical "control of chaos" problem is to transform $\tilde{x}(t)$ into a periodic trajectory $\bar{x}(t)$, $\bar{x}(0) = \bar{x}_0$, i.e., to satisfy the inequality

$$|x(t, u, x_0) - \bar{x}(t)| < \Delta_x \tag{6.7.50}$$

for sufficiently large t, where control $u = u(t)$ satisfies inequality

$$|u(t)| < \Delta_u. \tag{6.7.51}$$

As was previously mentioned, the goal trajectory $\bar{x}(t)$ may be either known, or not. The control action should be designed either as a function of time (feedforward or program control) or as a function of the state $x(t)$ (state feedback) or as a function of an output $y(t) = h(x(t))$ (output feedback).

Starting with the seminal paper [331] an *outer deadzone* approach is used by many authors. Its key point is that no control is applied if the current state if far from the target one, i.e., if the inequality (6.7.50) holds. Otherwise control function is calculated according to some stabilizing control law (e.g., linear feedback stabilizing Poincaré map in OGY or OPF methods). Obviously, for small Δ_x the constraint (6.7.51) will be valid too, if Δ_u and Δ_x are of the same order: $\Delta_x \sim l\Delta_u$.

To estimate the time required for control two terms should be taken into account: estimate of time t_1 required to achieve inequality (6.7.50) and estimate of remaining time t_2 when (6.7.50) holds. Since in the vicinity of the target point or trajectory the convergence rate is, in general, exponential, the time t_2 required to achieve the desired accuracy δ depends logarithmically on the ratio Δ_u/δ, i.e., $t_2 \sim \ln(\Delta_u/\delta)$. The main problem is to estimate t_1.

An estimate for t_1 was given by [331] who treated the first part of trajectory as a chaotic transient and used an estimate for wandering time of a chaotic trajectory approaching attractor when some parameter of the system approaches its critical bifurcation value [184]. It was noticed in [331] that t_1 depends sensitively on the initial condition and, if an initial condition is random, t_1 has exponential distribution. It means that the average time is $<t_1> \sim 1/p$, where p is the probability of entering in the set in the state space determined by the relation (6.7.50). In the case of a discrete-time two-dimensional system the power scaling was suggested for the probability: $p \sim \Delta_u^\gamma$, where $\gamma = 1 + 1/2 \ln|\lambda_u|/\ln|\lambda_s|^{-1}$.

Arguments explaining exponential distribution are rather simple. Let the system be described by discrete-time model $x_{t+1} = f(x_t, u_t)$ with sampling

6.7 Time and energy needed for control of chaos

interval equal to the desired cycle period, i.e., the goal for discrete-time model is to stabilize a fixed point $\bar{x}$. Then by ergodicity of the chaotic process the frequency of entering a set Ω during some time is proportional to the probability of the set and the length of the interval. Besides, if the probability p is small, the events consisting in entering the set are rare and can be considered as independent ones. Hence, the time t_Ω before entering the set Ω is distributed exponentially (geometrically in discrete time case) with probabilities $P\{t_\Omega = k\} = p(1-p)^k$, $k = 1, 2, \ldots$ and average $< t_\Omega > = 1/p$. Since $u(t) = 0$ outside the strip in the state space determined by inequality (6.7.51) and the probability p of the set Ω is proportional to its volume, the value of p is of order Δ_u, i.e., $p \sim \Delta_u$. For different shape of the deadzone, determined by the inequality (6.7.50), the value of p is of order $(\Delta_u)^n$, i.e., $p \sim (\Delta_u)^n$. Therefore in all cases the power law holds:

$$< t_1 > \sim (\Delta_u)^{-\gamma}. \tag{6.7.52}$$

It is possible to speed up control if we pass from passive to active control strategy between entries into Ω. Control intended to drive the system to the attractor faster is known as "targeting" or "directing," see [339, 340, 369]. Various control algorithms for the first stage of the process are proposed for special cases and their efficiency is demonstrated by means of simulation. However, analytical estimates are not presented. It is worth to note that the problem of global control of chaos is nonconvex (likewise the problem of global control of general nonlinear systems) and therefore hard to solve. Solutions that are good for special cases may lose efficiency for other cases.

In the paper [452], the power law for transient time T for OGY method was obtained $T \sim \frac{1}{\Delta_x}$, (it is easy to see that control level Δ_u is of the same order as Δ_x). Besides, in [452] a modification of OGY method was proposed for one-dimensional unimodal maps, possessing control time $T \sim \frac{2}{h^2} \log_2 \frac{1}{\Delta_u}$, where h is the entropy of the map.

A more general control algorithm was proposed in [229, 327]. It is based on symbolic dynamics approach and allows for analytical bounds of control time both for hyperbolic and for neutral systems to be obtained. Below a simplified version of the algorithm of [229, 327] will be described and some estimates for its control time for different classes of systems (maps) will be presented.

To start the algorithm a partition of a compact set D, containing both the goal attractor and the initial point into a finite number of cells $D_1, \ldots, D_N$ having sizes of order Δ_u should be found. The next step is to associate the system (6.7.49) with an oriented graph (called "skeleton" or symbolic image of the system). The vortex v_i corresponds to the cell D_i and the edge $v_i \to v_j$ exists if there exists a trajectory of (6.7.49) with $u(t) = 0$, starting within the cell D_i and leaving it through the cell D_j. Then the shortest route on the graph from the cell, containing the initial point x_0 to the cell, containing the end point x_f is sought. The movement between the cells of the chosen route is organized by means of "local" controls, providing smooth transitions between the neighbor cells. It is shown in [230, 231] that there exists control chosen by the above procedure, satisfying the bound (6.7.51) and providing the transient time which depends logarithmically on the cell size:

$$<t_1> \sim |\ln \Delta_u|. \qquad (6.7.53)$$

The main assumption allowing to prove (6.7.53) is hyperbolicity condition (exponential growth of the volumes along the trajectories of (6.7.49) with $u \equiv 0$). If hyperbolicity is replaced by conservation of volumes (neutral system), then the logarithmic law should be replaced by the power law [230, 231].

Though hyperbolicity is just a mathematical assumption, it describes a model class of systems exhibiting strong chaotic behavior. For example, Hamiltonian systems having both homoclinic and heteroclinic trajectories possess some properties of hyperbolic systems.

Another problem related to control of chaos is that of driving a trajectory from the initial point x_0 to the end point x_f, both within an attractor. Only driving with small control intensity, ideally with arbitrarily small control action is of interest in context of chaos control. Principal solvability of the problem follows from the fact that any chaotic trajectory is dense in its closure coinciding with the whole attractor. Similarly, it can be verified that if control is switched off during traveling along a chaotic trajectory, then the estimate of the required time has the power order (6.7.52).

A more efficient algorithm for this problem was proposed in [230, 231]. As in the previous case it consists of three stages. At the first stage the trajectory should be driven to a prespecified dense chaotic trajectory of the unforced system (local control). At the second stage, using symbolic dynamics approach a sequence of piecewise smooth control actions is chosen driving the trajectory according to the shortest route in the skeleton graph to a vicinity of the final point. It is shown in [230, 231] that under hyperbolicity assumption the transport time depends logarithmically on the control intensity Δ_u, see (6.7.53), while for neutral systems logarithmic law should be replaced by the power one.

The above estimates of the time required for control have interesting consequences for estimation of required energy. If the control action has physical nature of force or torque, then to obtain power of control it should be multiplied by the output which is bounded. Therefore, the power of control has the order Δ_u and the energy has the order $\Delta_u(t_1+t_2)$. Based upon the algorithm of [230, 231] for hyperbolic case the estimate of energy E has the order

$$E \sim \Delta_u |\ln \Delta_u|. \qquad (6.7.54)$$

It follows from (6.7.54) that choosing small Δ_u control of a chaotic trajectory can be achieved with arbitrarily small energy expenditure. Note that it is not the case for neutral systems where only the required power can be made arbitrarily small.

6.8 Applications in physics

Below a number of applications of chaos control in various fields of physics will be surveyed briefly. Time and space limits do not allow us to discuss many other interesting applications, e.g. in electronics [322], physical chemistry [68, 232, 346], semiconductors [50, 409], etc.

6.8.1 Control of turbulence

Description and control of turbulence remains one of the main physical problems already over an entire century [171]. The infinite-dimensional description of the turbulent flow as a solution of the Navier–Stokes partial equation is known to be often reducible to the finite-dimensional description. If the dimensionality of the flow attractor in the phase space is relatively small, then the turbulent flow may be regarded as chaotic, and methods of chaos control can be applied to it. The Taylor–Couette flow of liquid between two rotating concentric cylinders exemplifies such flows.

The approximating dimension in wall-bounded flows like turbulent Poiseuille flows appears to be rather high – typical values are over several hundred. In contrast in closed, absolutely unstable flows, such as Taylor–Couette systems, the number of degrees of freedom can be small [171]. Recently, applicability of chaos control method to such a class of turbulent motion was confirmed experimentally.

Experimental control of the dynamics of the chaotic structures arising in the Taylor vortex flow with globoid (hourglass) geometry is described in [447]. This flow is a variant of the Taylor–Couett system. In the experiment, the internal cylinder was rotated by a computer-controlled step motor. A water–glycerol mix with 1.5 volume percent of the Kalliroscope suspension added for visualization was used. An increase in the rotation velocity leads to a greater Reynolds number $R = 2\pi f a d/\nu$, where f is the rotation frequency, a is the globoid radius at the central part, d is the gap at the center, and ν is the kinematic viscosity. For $R > R_{ps}$, where R_{ps} is the critical Reynolds number for which phase slip occurs, pairs of vortices arise: first, periodic, then chaotic. The intervals I_n between phase slips were measured by a TV camera. Control was exercised by varying the reduced Reynolds number $\varepsilon = (R/R_{ps}) - 1 = (f/f_{ps}) - 1$ by means of the algorithm

$$\delta\varepsilon_{n+1} = K(I_n - I_F) + R\delta\varepsilon_n, \qquad (6.8.55)$$

where $\delta\varepsilon_n = \varepsilon_n - \bar{\varepsilon}$ and $\bar{\varepsilon}$ corresponds to the periodic unstable motion with the interval I_F between the phase bands. The control signal was fed only if the condition $|\delta\varepsilon_{n+1}| < 0.01$ was satisfied, which corresponds to the generalized OPF-algorithm (see Section 6.4.2). The parameters I_F, K, and R of the control law were selected experimentally. It was established that in order to suppress chaos for $\bar{\varepsilon} = 0.417$ (corresponding to the chaotic process) it suffices to vary ε at most by 2 %. Luthje et al. demonstrated that similar results can be obtained by time-delayed feedback [279].

6.8.2 Control of lasers

Suppression of chaotic (so-called *multimode*) behavior in lasers was one of the first reported applications in the field [381]. This work presented experimental data on feedback leading which enabled a substantial (by an order of magnitude) increase in the radiation power owing to more powerful pumping. In 1997–2000 a few dozen journal papers related to the control of chaos in laser and optical devices were published. Recent research has focused on attempting to control a range of instabilities in different types of lasers by a variety of methods. The most common methods are feedforward and delayed feedback. Experimental suppression of Lorenz-like instability in ammonia lasers by delayed feedback was demonstrated in [124].

The methods of control by open loop and time-delayed feedback for the CO_2-lasers with modulation of losses and also for the doped Nd fiber laser were compared in [179] which numerically predicted for the lasers of the class B that their stability domain would be extended (shift of the period duplication bifurcation) for the control based on the models with two degrees of freedom. Analytical facts were corroborated by models and experiments.

6.8.3 Control of chaos in plasma

A number of papers reported successful control of chaos in the so-called *Pierce diode* [166, 228, 348]. A Pierce diode is the simplest model to analyze the stability of current flow in a plasma diode, where both virtual kinetic cathode oscillations and hydrodynamic plasma oscillations appear. The OGY method was used in [348] to stabilize cycles of periods 1 and 2. The signal of time-delayed feedback by measurements of the space charge density at a fixed space point was used in [228] to suppress chaos by modulating the difference of potentials between the input and output diode grids. The results can be applied to bring the Pierce diode into a well-defined state of microwave oscillations.

Recent results on multimode feedback control of magnetohydrodynamic modes and a variety of diagnostic uses of feedback in plasma were summarized in [397]. Their primary goal is an experimental methodology for the determination of dynamic models of plasma turbulence, both for better transport understanding and more credible feedback controller designs. In [397], a new method for direct experimental determination of nonlinear dynamical models of plasma turbulence using feedback is reported. The results are confirmed experimentally in the Columbia Linear Machine and can be extrapolated to fusion machines as neutron beam suppressors.

Interaction of the laser radiation with plasma which plays an important role in the problems of controlled thermonuclear fusion was studied in [382]. It is noted that on the whole the process of this interaction is very complicated and its mathematical model has not yet been obtained, but it is known that two types of phenomena are observed at interaction. One lies in occurrence of stable soliton-like structures, and the other, in occurrence of extremely unstable chaotic processes. It is assumed that these phenomena can be studied separately. The authors of [382] examine the means of reducing the chaotic process to periodic oscillations or steady state, respectively, by the controls of two types: by open-loop periodic variation of a system parameter or by the so-called "proportional pulse control." The paper focuses on the following model:

$$\begin{cases} \dot{x} = -gx - b_1(x+z)y^2, \\ \dot{y} = -g_0 y + b_2(x^2 - z^2)y, \\ \dot{z} = -g_0 z + b_3(x+z)y^2, \end{cases}$$

where x, y, and z are, respectively, the dimensionless field amplitudes for anti-Stokes, pump, and Stokes modes and g, g_0, b_1, b_2, and b_3 are the parameters. Differentiation is carried out with respect to the spatial coordinate along the direction of wave propagation. The open-loop control lies in varying the parameter g_0 according to $g_0(t) = \bar{g}_0 - a\cos\omega_0 t$. For $\bar{g}_0 = 1$ and $a = 0$ (no control), chaotic behavior is observed. The system process can be driven to periodic by an appropriate choice of the amplitude a and frequency ω_0 of modulation of the parameter g_0. Proportional pulse control lies in discontinuous variation of the state variables at certain time instants.

The results of experimental studies of the chaotic processes in n-conductivity germanium oscillators can be found in [205]. Behavior of Kadomtsev–Nedopasov instability in electron-hole plasma at temperatures 77°K and 300°K under the action of external electrical and magnetic fields was studied. Pictures of space-time development of chaotic processes were obtained from the measurements at various points of specimens. Bifurcation diagrams showing the boundaries of domains with double period, quasiperiodicity, chaoticity, and intermittence were obtained. Several attractors having each its own dimension and energy response were shown to be feasible simultaneously in specimens for certain conditions.

Synchronization of the chaotic space-time structures (patterns) in the spatially distributed models of semiconductor heterostructures by means of the time-delayed feedback was considered in [409] where control with the diagonal feedback matrix, global control, and their combination were compared. Consideration was given to two models of semiconductor nanostructures that are of current interest: superlattice and two-barrier diode with resonance tunneling. Quality of control in these systems was shown [16] to improve by several orders of magnitude owing to suitable filters and couplings based on the Floquet eigenmodes of the unstable orbits. For the mechanism resulting in a better control on the basis of phase synchronization of the desired process and that in the control loop, an explanation was given.

6.9 Other problems

From the problems relating to control of chaos, the following ones remain the most important.

Controllability. Although controllability of nonlinear systems is well studied, few results are available on reachability of typical control goals by small control [11, 152]. Among basic tools are Pugh lemma and Anosov lemma, see Section 6.2. A general idea, illustrated by many case studies is that the more a system is "unstable" (chaotic, turbulent) the "simpler," it is to achieve its exact or approximate controllability. In other words, "chaos facilitates control" [84, 273, 435].

Other control goals. Among alternative control goals such as achieving the desired period [133], the desired fractal dimension of the attractor [371] the desired invariant measure [84, 183] desired Kolmogorov entropy [336] are studied in the literature. A method for the so-called *tracking chaos* problem: following a time-varying unstable orbit, based on the continuation method for solving equations was proposed by I. Schwartz and I. Triandaf [394] as early as in 1992.

Chaos in control systems. Control of chaos should not be mixed up with *chaos in control systems*. The results in the latter area, providing conditions for chaotic behavior in conventional feedback control systems were published since the 1970s [51, 281, 284]. They are related to chaos in linear delayed systems [281], nonlinear systems [51], adaptive systems [284]. In the recent publications conditions for chaotic behavior are provided for mechanical control systems [15, 127], systems with hysteresis [357], systems with pulse-width modulation [233], to mention a few.

Conclusions. Apparently, the publication rate in control of chaos has achieved saturation by the end of the 20th century, see Fig. 1.2.1. However, it is seen from the statistics that the field is again rapidly developing in the beginning of the 21st century. Today, there are many efficient methods for control of chaos in the literature. Three major and historically the first schemes are feedforward ("nonfeedback") control, the OGY method and the Pyragas method. Using the methods of nonlinear and adaptive control is very promising. However, special care should be taken to respect "small control" requirement. Some important problems of justification of existing methods remain unsolved and provide challenges both for physicists and for control theorists.

7

Control of Interconnected and Distributed Systems

7.1 Models of controlled spatiotemporal systems

Among infinite-dimensional (distributed) systems the main classes are spatially extended (spatiotemporal) systems and retarded or delayed systems. Methods of oscillations and chaos control in infinite-dimensional systems are mainly based upon ideas developed for finite-dimensional (lumped) systems. Moreover, finite-dimensional models are often used for control system analysis and design.

Finite-dimensional models of spatially extended controlled systems are obtained by spatial discretization of distributed models described by partial differential equations (PDE). Such simplified models consist of ordinary differential equations (ODE) describing separate space elements called cells, particles, compartments. In both cases elements (cells) interact by means of links reflecting the spatial structure of the overall system (e.g., array, lattice).

A broad class of controlled spatiotemporal systems are described by controlled *reaction–diffusion* equations

$$\frac{\partial x}{\partial t} = \varepsilon \Delta x + F(x, u), \tag{7.1.1}$$

where $x = x(r, t)$ is a function of space variables $r \in D \subset \mathbb{R}^n$ and time t (possibly, vector-valued), determining the state of a physical system, $\Delta = \sum_{i=1}^{n} \frac{\partial^2}{\partial r_i}$ is Laplace operator, specifying diffusion type of space elements interaction, ε is diffusion coefficient. Boundary conditions can be specified as periodic ones (e.g., $x(a,t) = x(b,t)$ for $D = [a,b] \subset \mathbb{R}^1$) or "no flow across the boundary" conditions $\left(\frac{\partial x}{\partial r}\right)\big|_{r=a} = \left(\frac{\partial x}{\partial r}\right)\big|_{r=b} = 0$.

For $\varepsilon = 0$ equation (7.1.1) reduces to a ODE which may exhibit chaotic behavior. A natural approach to studying equation (7.1.1) is its approximation by a set of ODE by means of discretizing over the space D. It means replacing continuum D by a finite number of points (nodes) r_i, $i = 1, 2, \ldots, N$.

The dynamics of each state x_i depends on both its internal (local) dynamics $F(x_i, u_i)$ and interactions with neighbor nodes. For example, if the space variable is one-dimensional: $r \in [a, b]$, and interactions are of diffusive type, then the space-discretized model has the form

$$\dot{x}_i = \varepsilon(x_{i-1} - 2x_i + x_{i+1}) + F(x_i, u_i), \quad i = 1, 2, \ldots, N - 1. \quad (7.1.2)$$

Additionally, boundary conditions should be specified, e.g., as periodic ($x_0 = x_N(t)$, or "no-flow" $x_0(t) = x_1(t)$, $x_{N-1}(t) = x_N(t)$. Sometimes the models that are discretized both in space and in time are used. They are called *coupled maps* or *cellular automata* models:

$$x_i(n+1) = x_i(n) + \varepsilon \left[x_{i-1}(n) - 2x_i(n) + x_{i+1}(n)\right] + hF(x_i(n), u_i(n)), \quad (7.1.3)$$
$$i = 1, \ldots, N - 1, \quad n = 0, 1, 2, \ldots$$

It can be seen that control in the models (7.1.1), (7.1.3) influences the dynamics of each cell that corresponds to the case of space-distributed (field) control. Another class of boundary control problems arises when right-hand sides in (7.1.2)–(7.1.3) do not depend on control, i.e., $F(x, u) \equiv F(x)$, while control enters only equations of boundary cells, e.g.,

$$\dot{x}_0 = \varepsilon(x_1 - x_0) + F_0(x, u), \quad (7.1.4)$$

The situation may be further generalized to consider space-nonhomogenious systems. For the one-dimensional case they are described by the following model:

$$\begin{aligned} \dot{x}_i &= F_i(x_i, x_{i-1}, x_{i+1}, u), \quad i = 1, 2, \ldots, N-1, \\ \dot{x}_0 &= F_0(x_0, x_1, u), \\ \dot{x}_N &= F_N(x_N, x_{N-1}, u). \end{aligned} \quad (7.1.5)$$

Control goals can be straightforward extensions of the goals formulated for lumped systems (see Section 2.2). In addition, specific goals can be posed formalizing a specific type of interrelation between neighbor cells.

Among specific spatiotemporal control goals the following ones should be mentioned:

– stabilization of the given uniform (homogeneous) or space-periodic field (standing wave) – such goals are well studied in conventional control theory [93];
– stabilization of the given time-periodic motion (traveling wave);
– creation or suppression of spiral wave (for space dimension not less than two);
– creation or suppression of the given nonhomogenious field (contrast or dissipative structure, clusters, patterns);
– control of self-organization or disorganization of systems.

7.2 Control of energy in sin-Gordon and Frenkel–Kontorova models

Control problems for distributed systems were systematically studied as early as in the 1960s [93]. However the interest of physicists has become apparent as late as in the middle of the 1990s. Perhaps, it was motivated by the interest in control of chaos in distributed systems. A brief survey of results in control of chaos in distributed systems will be presented in Section 7.5.

7.2 Control of energy in sin-Gordon and Frenkel–Kontorova models

Let us study possibilities of the speed-gradient method for control of energy in sin-Gordon-like systems. Using the following notations: $x_t = \frac{\partial x}{\partial t}$, $x_{tt} = \frac{\partial^2 x}{\partial t^2}$, $x_{r_i} = \frac{\partial x}{\partial r_i}$, $x_{r_i t} = \frac{\partial^2 x}{\partial r_i \partial t}$, $x_{r_i r_i} = \frac{\partial^2 x}{\partial r_i^2}$, consider the system, described by sin-Gordon equation with dissipation

$$J x_{tt} = k \Delta x - E \sin x - \rho x_t, \qquad (7.2.6)$$

where $x = x(r,t)$ is the function of the system state; $r \in X \subset \mathbb{R}^n$ is the spatial variable, taking values from a set X; Δ is Laplace operator; $\Delta x = \sum_{i=1}^{n} x_{r_i r_i}$; J, k, ρ are parameters of the system; $E = E(t)$ is the external action (e.g., external force or intensity of the external electrical field). Assume that $E = E_0 + u(t)$, where E_0 is the base level of the intensity of the force or field; $u(t)$ is the controlling variable. The system (7.2.6) can be considered as a model of diffusively coupled oscillators (e.g., pendulums, magnetic domains liquid crystals), each being positioned in the spatial point r. Then $x(r,t)$ is the deflection angle of rth oscillator at time t. Such a system belongs to a class of reaction-diffusion systems, but its study is of independent value.

Let us pose the problem of controlling the energy of the free system

$$H = \frac{1}{2} \int_X \left(J \left(\frac{\partial x}{\partial t}\right)^2 + k |\nabla_r x|^2 + 2 E_0 (1 - \cos x) \right) dr \qquad (7.2.7)$$

to the prespecified level H_*. It means that we introduce the control goal as follows:

$$\lim_{t \to \infty} H(t) = H_*. \qquad (7.2.8)$$

First, let $\rho = 0$ and evaluate the rate of changing the energy along trajectories of the system (7.2.6) assuming that the controlling variable is frozen: $u(t) = u$:

$$\begin{aligned}
\frac{dH}{dt} &= \int_X J x_t \cdot x_{tt} - k \Delta x x_t + E_0 \sin x \cdot x_t \, dr \\
&= \int_X x_t \left(-E \sin x + E_0 \sin x \right) dr = -u(t) \int_X x_t \sin x \, dr.
\end{aligned} \qquad (7.2.9)$$

It is easy to see that the choice of the control in the form

$$u(t) = -\gamma \int_X x_t \sin x \, dr, \qquad (7.2.10)$$

where $\gamma > 0$, guarantees that the energy $H(t)$ will not decrease in time.

Introducing the goal function $V(t) = \frac{1}{2}(H(t) - H_*)^2$, and evaluating the time derivative of $V(t)$, we obtain

$$\dot{V} = \frac{dV}{dt} = -u(t)\big(H(t) - H_*\big) \int_X x_t \sin x \, dr, \qquad (7.2.11)$$

and $\dot{V} \leq 0$ for

$$u(t) = \gamma\big(H(t) - H_*\big) \int_X x_t \sin x \, dr. \qquad (7.2.12)$$

Thus, if the system is affected by the action (7.2.12) it will have a tendency to approach the goal.

Consider in more detail the spatially one-dimensional, spatially discrete version of the problem, described by equations

$$J\ddot{x}_j = \frac{k}{h^2}\big(x_{j+1} - 2x_j + x_{j-1}\big) - \big(E_0 + u(t)\big) \sin x_j - \rho \dot{x}_j, \quad j = 1, 2, \ldots, N. \qquad (7.2.13)$$

It corresponds to a continuous system

$$J x_{tt} = k x_{rr} - \big(E_0 + u(t)\big) \sin x - \rho x_t \qquad (7.2.14)$$

defined in the set $X = [a, b]$, if the correspondence is defined by the relations $x_j = x(a + j(b-a)/(N+1)), j = 0, 2, \ldots, N+1$.

The system (7.2.13) is suggested to be a controlled version of the classical *Frenkel–Kontorova chain*, proposed in 1939 and studied in numerous works, see, e.g., [253].

Before designing the control law, let us discuss the choice of boundary conditions. Usually when studying an uncontrolled system (7.2.14) two types of boundary conditions are used: either zero boundary conditions $x(a, t) = x(b, t) = 0$, corresponding in the discrete system (7.2.13) to the relations

$$x_0(t) \equiv x_{N+1}(t) \equiv 0, \qquad (7.2.15)$$

or periodic (no flux across the boundary) conditions $x_r\big|_{r=a} = x_r\big|_{r=b} = 0$, corresponding to the relations

$$x_0 = x_1, \qquad x_N = x_{N+1}. \qquad (7.2.16)$$

The control problem for energy control of the chain can be solved based on the results of the Chapter 3. The speed-gradient energy control algorithm looks as follows:

7.2 Control of energy in sin-Gordon and Frenkel–Kontorova models

$$u(t) = \gamma \big(H(t) - H_*\big) \sum_{j=1}^{N} \dot{x}_j \sin x_j, \qquad (7.2.17)$$

where $\gamma > 0$. It follows from Theorem 3.1 that the control goal (7.2.8) in the system (7.2.13), (7.2.17) for $\varrho = 0$ is achieved if the energy layer in the system phase space between the energy levels $H(0)$ and H_* does not contain equilibria satisfying conditions $\sin x_j = 0, j = 1, \ldots, N$.

For the special case $N = 2$ with the boundary conditions (7.2.16) the system takes the form

$$\begin{cases} J\ddot{x}_1 = \dfrac{2k}{h^2}(x_2 - x_1) - \big(E_0 + u(t)\big)\sin x_1 - \rho \dot{x}_1, \\ J\ddot{x}_2 = \dfrac{2k}{h^2}(x_1 - x_2) - \big(E_0 + u(t)\big)\sin x_2 - \rho \dot{x}_2. \end{cases} \qquad (7.2.18)$$

For this special case the energy control problem is close to the energy/synchronization control problem examined in Section 5.4. The discretized version of the control algorithm, corresponding to (7.2.12) is as follows:

$$u(t) = \gamma \big(H(t) - H_*\big)(x_2 - x_1)\big(\sin x_1 - \sin x_2\big), \qquad (7.2.19)$$

where $H(t)$ has discretized form (7.2.7)

$$H = \frac{J}{2}(\dot{x}_1^2 + \dot{x}_2^2) + \frac{k}{2}(x_1 - x_2)^2 + E_0\big(2 - \cos x_1 - \cos x_2\big). \qquad (7.2.20)$$

The proposed algorithms can be used to explore oscillatory properties of nonlinear systems in various problems. For example, they can be used to control orientation of the oscillating particles, e.g., liquid crystals, where the control goal is to change the orientation from longitudinal to transverse or vice versa, see Fig. 7.2.1.

Suppose that the orientation angles $x(r, t)$ of the particles obey the equation (7.2.6). Then it is possible to transform the longitudinally oriented structure (a) into the transversely oriented structure (b) by means of changing the orientation (polarization) of the external constant field E. However, for this purpose the required intensity of the field may be significant. At least its change δE should exceed the intensity of the initial field E_0.

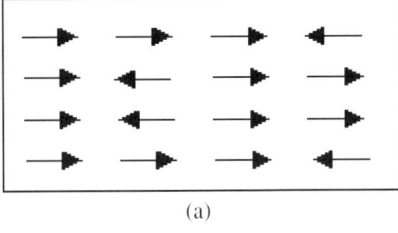
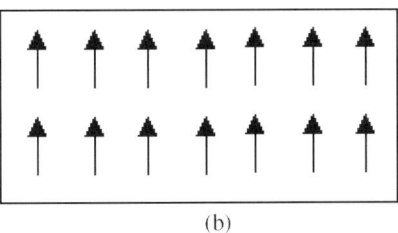

Fig. 7.2.1. Nontransparent (a) and transparent (b) structures of liquid crystal array.

At the same time using a feedback control algorithm (7.2.12) makes the particles perform an oscillatory pendulum-like motion, and may significantly reduce the intensity required for control. To this end the desired energy H_* is chosen as the value close to energy of the configuration b, i.e., $H_* = \bar{H} = E_0(b-a)$. Then, if the dissipation $\varrho > 0$ is sufficiently small, the algorithm (7.2.12) makes the particles spend most part of their time near the configuration b, i.e., makes the crystal highly transparent. According to the results of Chapter 4 (Theorem 4.1) the required controlling field intensity γ should be of order $\varrho\sqrt{H_*}$, i.e., it may be small provided the dissipation $\varrho > 0$ is small.

7.3 Control of wave motion in the chain of pendulums

7.3.1 Modeling the chain of the pendulums

Consider, following [111], the problem of controlled excitation of oscillations in the chain of N coupled mathematical pendulums. Such a model may arise when studying different physical and mechanical systems, see, e.g., [211, 253]). In the absence of the friction the system of coupled pendulums is described by the equations

$$\begin{cases} \ddot{\varphi}_1(t) + \omega_0^2 \sin\varphi_1(t) = k(\varphi_2(t) - \varphi_1(t)) + u(t), \\ \cdots\cdots\cdots\cdots\cdots\cdots\cdots\cdots\cdots\cdots\cdots\cdots \\ \ddot{\varphi}_i(t) + \omega_0^2 \sin\varphi_i(t) = k(\varphi_{i+1}(t) - 2\varphi_i(t) + \varphi_{i+1}(t)), \\ \qquad\qquad\qquad (i = 2, 3, \ldots, N-1), \\ \cdots\cdots\cdots\cdots\cdots\cdots\cdots\cdots\cdots\cdots\cdots\cdots \\ \ddot{\varphi}_N(t) + \omega_0^2 \sin\varphi_N(t) = k(\varphi_{N-1}(t) - \varphi_N(t)), \end{cases} \qquad (7.3.21)$$

where $\varphi_i(t)$ $(i = 1, 2, \ldots, N)$ are the pendulum deflection angles; $u(t)$ is the controlling action: external torque, applied to the first pendulum. It is assumed that the torque is measured in the units of angular acceleration. The values ω, k are parameters of the system: ω_0 is the natural frequency of small oscillations of isolated pendulums, k is the parameter of coupling strength, for example, stiffness of the spring connecting the pendulums.

Introduce the state vector of the system $x(t) \in \mathbb{R}^{2N}$ as follows: $x(t) = \mathrm{col}\{\varphi_1, \dot{\varphi}_1, \varphi_2, \dot{\varphi}_2, \ldots, \varphi_N, \dot{\varphi}_N\}$. The total energy of the system (7.3.21) $H(x)$ is defined by the expression

$$H(x) = \sum_{i=1}^{N} H_i(x), \quad \text{where}$$

$$\begin{cases} H_i(x) = 0.5\dot{\varphi}_i^2 + \omega_0^2(1 - \cos\varphi_i) + 0.5\,k(\varphi_{i+1} - \varphi_i)^2 \\ \qquad\qquad\qquad (i = 1, 2, \ldots, N-1), \\ H_N(x) = 0.5\dot{\varphi}_1^2 + \omega_0^2(1 - \cos\varphi_N). \end{cases} \qquad (7.3.22)$$

7.3 Control of wave motion in the chain of pendulums

In the absence of control the model in question coincides with the Frenkel–Kontorova model described in the previous section, with the only exclusion that the friction is neglected. However, it is distinct from the Frenkel–Kontorova model in the type of the control appearance: here control is localized and affects only one pendulum. In terms of control of distributed systems it corresponds to boundary control.

In addition, we will consider the system of cyclically coupled pendulums, similar to (7.3.21), except the presence of the elastic link between the first and the last pendulums:

$$\begin{cases} \ddot{\varphi}_1(t)+\omega_0^2 \sin \varphi_1(t)=k\big(\varphi_2(t)-2\varphi_1(t)+\varphi_N(t)\big)+u(t), \\ \quad \dotfill \\ \ddot{\varphi}_i(t)+\omega_0^2 \sin \varphi_i(t)=k\big(\varphi_{i+1}(t)-2\varphi_i(t)+\varphi_{i+1}(t)\big) \\ \quad (i=2,3,\ldots,N-1), \\ \quad \dotfill \\ \ddot{\varphi}_N(t)+\omega_0^2 \sin \varphi_N(t)=k\big(\varphi_{N-1}(t)-2\varphi_N(t)\big)+\varphi_1(t)). \end{cases} \quad (7.3.23)$$

The expression for the total energy changes correspondingly:

$$H(x)=\sum_{i=1}^{N} H_i(x), \quad \text{where}$$

$$\begin{cases} H_i(x)=0.5\,\dot{\varphi}_i^2+\omega_0^2(1-\cos\varphi_i)+0.5\,k\big(\varphi_{i+1}-\varphi_i\big)^2 \\ \quad (i=1,2,\ldots,N-1), \\ H_N(x)=0.5\,\dot{\varphi}_1^2+\omega_0^2(1-\cos\varphi_N)+0.5\,k\big(\varphi_1-\varphi_N\big)^2. \end{cases} \quad (7.3.24)$$

The equations (7.3.23) are symmetrical with respect to the free motion of the pendulums. Such a symmetry allows the achievement of an additional control goal: synchronization of the pendulum motions.

7.3.2 Problem statement and control algorithm design

Let us interpret the problem of the excitation of a wave as achievement of the given level of the system energy with additional requirement that the neighbor pendulums have opposite oscillation phases. Let us use the speed-gradient method for the control algorithm design.

To apply the speed-gradient method introduce two auxiliary goal functions.

$$\begin{aligned} Q_\varphi(\dot{\varphi}_1,\dot{\varphi}_2) &= 0.5\,\delta_\varphi^2, \\ Q_H(x) &= 0.5(H(x)-H^*)^2, \end{aligned} \quad (7.3.25)$$

where $\delta_\varphi = \dot{\varphi}_1 + \dot{\varphi}_2$; $H(x(t))$ is *the total energy of the system*; H^* is the desired value of energy.

Apparently, the minimum value of the function Q_φ corresponds to the anti-phase motion of the first and second pendulums since the identity

$Q_\varphi(\dot\varphi_1, \dot\varphi_2) \equiv 0$ holds only if $(\dot\varphi_1 \equiv -\dot\varphi_2)$. The minimization of Q_H corresponds to achievement of the desired oscillation amplitude.

Let us introduce the total goal function $Q(x)$ as the weighted sum of Q_φ and Q_H, namely

$$Q(x) = \alpha Q_\varphi(\dot\varphi_1, \dot\varphi_2) + (1-\alpha) Q_H(x), \qquad (7.3.26)$$

where α $(0 \le \alpha \le 1)$ is the weighting coefficient (design parameter).

Evaluation of the speed of changing the goal function $Q(x)$ along trajectories of the controlled system and then its partial derivative in control leads to the following speed-gradient algorithm in finite form

$$\begin{aligned} u(t) &= -\gamma \left(\alpha \delta_\varphi(t) + (1-\alpha) \delta_H(t) \dot\varphi_1(t) \right), \\ \delta_\varphi(t) &= \dot\varphi_1(t) + \dot\varphi_2(t), \\ \delta_H(t) &= H_t - H^*. \end{aligned} \qquad (7.3.27)$$

Note that calculation of the control action according to the equation (7.3.27) requires the possibility of the measurement of the angular velocities of the first and second pendulums, as well as the measurement of the total energy of the system.

7.3.3 Simulation results

In Figs. 7.3.2–7.3.10 the results of computer simulations for the process of oscillations excitation by means of control law (7.3.27) for the chain of $N = 50$ pendulums are presented.

Figures 7.3.2–7.3.8 are related to the control of the system (7.3.21) by the control algorithm (7.3.27) for $\gamma = 0.8$ and various values of weighting coefficient α. Figures 7.3.2–7.3.4 correspond to the value $\alpha = 0$, when the control goal is stabilization of the given total energy level.

It is seen that the control goal $H_t \to H^*$ $(H^* = 4)$ is achieved but the motion of the pendulums is irregular (chaotic) and the second goal (synchronization) is not achieved, see Fig. 7.3.3. It is seen also that the control intensity decays as soon as the goal is achieved.

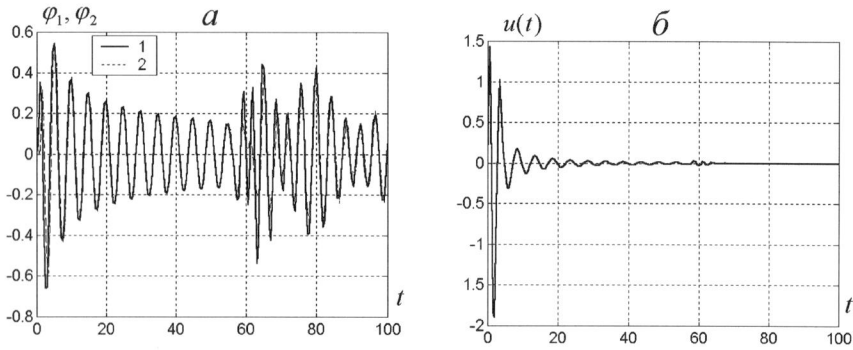

Fig. 7.3.2. Transient processes in φ_1, φ_2 and the control signal. $N=50$, $\alpha=0$.

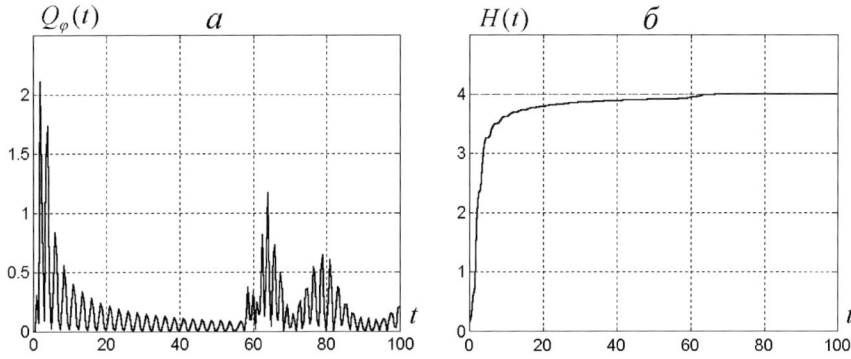

Fig. 7.3.3. The goal functions Q_φ and H_t. $N=50$, $\alpha=0$.

The reason of the irregular behavior is clear from Fig. 7.3.4. It is seen that the forward wave of oscillations propagates ahead and reaches the last pendulum at the time $t \approx 30$. Then the backward wave arises and after some time the picture of oscillations becomes complicated owing to interference of the waves. Excitation of oscillations for $\alpha = 1$ is shown in Figs. 7.3.5, 7.3.6. Again, it is seen that the antiphase motion of the pendulum is not achieved. For the choices of intermediate values of the weighting coefficient, $0 < \alpha < 1$, the behavior of the process is qualitatively the same as in Figs. 7.3.2–7.3.4 and the goal is not achieved.

A possible explanation of the control failure with respect to the goal function Q_φ is in that the antiphase free motion may not invariant, i.e., for $u(t) \equiv 0$ such a motion, if it arises once, may be destroyed in future. To verify this hypothesis consider Fig. 7.3.7 which is showing free oscillations in the system of $N = 50$ pendulums at anti-phase initial conditions ($\varphi_{i+1}(0) = -\varphi_i(0)$, $i = 1, 2, \ldots, N-1$). One can see the boundary effects of wave reflection leading to corruption of smooth oscillations after some time. It is interesting that the first pendulums to exhibit corruption of the wave (dislocation) are the first and the last ones.

The reason of appearance of the reflected wave is in the symmetry breaking of equations (7.3.21): the elasticity forces acting on the first and the last pendulums differ from the force acting on the internal pendulum in the chain.

Let us examine the system of cyclically coupled pendulums (7.3.23). It is seen from Fig. 7.3.8 that such a system exhibits no boundary effects. Note that the initial conditions for Fig. 7.3.8 coincide with those for Fig. 7.3.7.

The previous considerations justify that the controlled system of cyclically coupled pendulums may possess antiphase synchronous behavior at the specified energy level. The results of the simulation confirm such a conjecture. Figuress. 7.3.9, 7.3.10 demonstrate the results of application of the algorithm (7.3.27) to the system (7.3.23). The algorithm design parameters are as follows: $\gamma = 0.8$, $\alpha = 0.7$. It is seen, that the combined goal function tends to its minimum value.

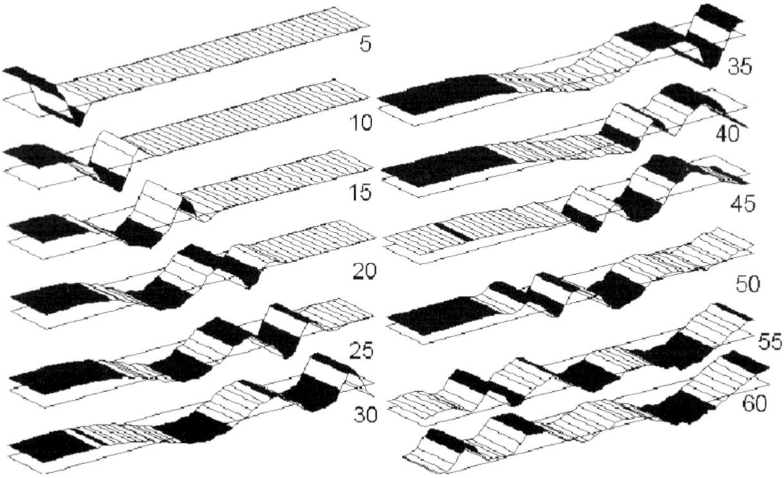

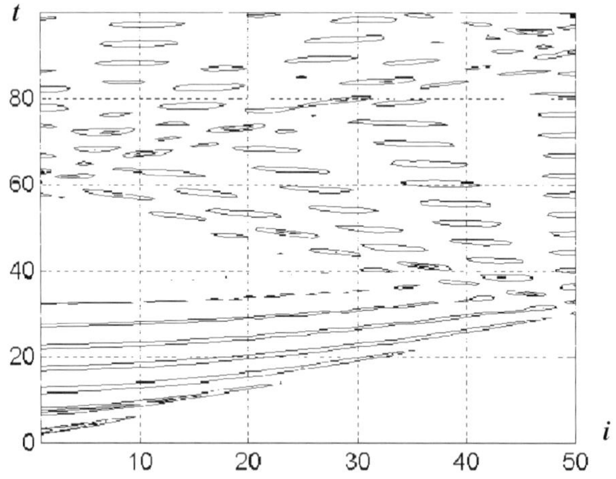

146   7 Control of Interconnected and Distributed Systems

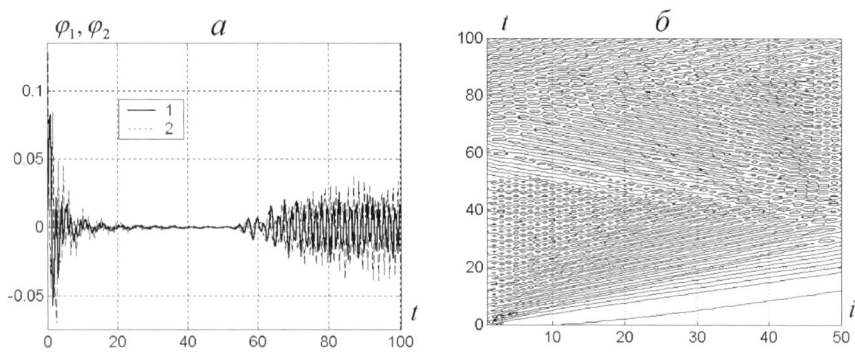

Fig. 7.3.4. Wave of oscillations. $N=50$, $\alpha=0$.

Fig. 7.3.5. Excitaton of oscillations. $N=50$, $\alpha=1$.

7.3 Control of wave motion in the chain of pendulums 147

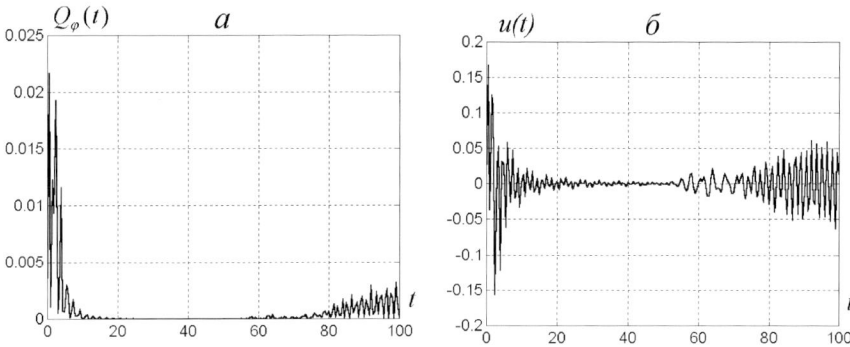

Fig. 7.3.6. Synchronization goal function Q_φ and control signal. $N=50$, $\alpha=1$.

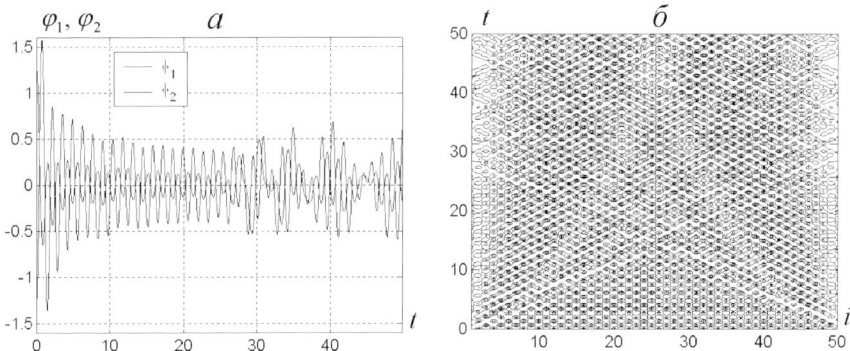

Fig. 7.3.7. Free oscillation of the chain. $N=50$, $u(t) \equiv 0$.

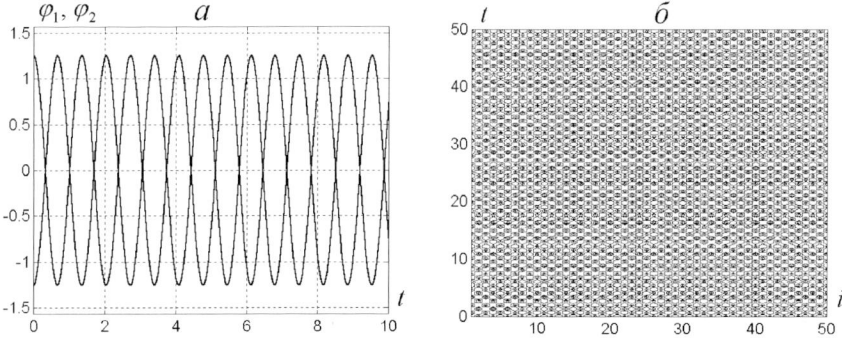

Fig. 7.3.8. Free oscillations of the cyclic chain. $N=50$.

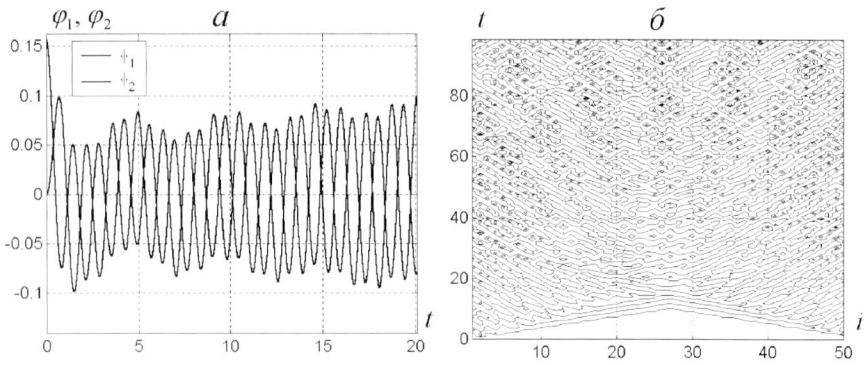

Fig. 7.3.9. Control of oscillation excitation for cyclic chain.

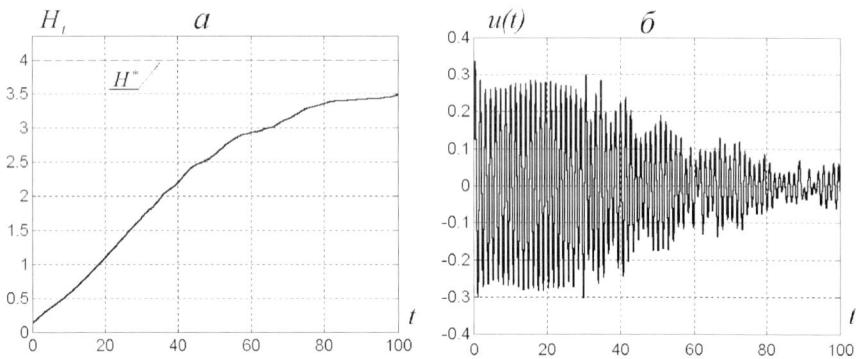

Fig. 7.3.10. Total energy and control action for cyclic chain.

7.4 Control of oscillations in a complex crystalline lattice

Properties of complex oscillatory systems such as atomic lattices are determined by the interaction of the large number of degrees of freedom. Studying such properties as structure and phase transitions, formation of defects, shock waves, requires consideration of strongly nonlinear phenomena. Some nonlinear effects may arise spontaneously and can be studied based on free oscillations theory. However, purposeful changes of the crystal state demand for development of the methods of controlling its properties, particularly, control of its nonlinear oscillations.

In this section, two models of complex crystalline lattice are described and the problem of excitation of oscillations is posed. The speed-gradient control algorithm is developed and the dynamics of the closed-loop system is analyzed. Presentation is based on the results of [5, 6].

7.4.1 Modeling interaction of acoustic and optical modes

In the paper by E. Aero [4] the following system of equations describing interaction of acoustic and optical modes in a nonlinear continuum model of crystal without center of symmetry (Aero model) is proposed:

$$\rho \ddot{U}_i = c_{ikj} u_{k,j} + \lambda_{ikjm} U_{k,jm};$$
$$(.)_{k,j} \to \partial(.)_k/\partial x_j; \quad (.)_{k,jn} \to \partial^2(.)_k/\partial x_j \partial x_n \qquad (7.4.28)$$

$$\mu \ddot{u}_i = -\frac{\partial \Phi}{\partial u_i} - \bar{c}_{kij} u_{k,j} + \kappa_{ikjm} u_{k,jm}. \qquad (7.4.29)$$

Here, $U_i(x_1, x_2, x_3, t)$, $u_i(x_1, x_2, x_3, t)$ are unknown functions describing the components of the displacements due to acoustic and optical modes in crystal, respectively. The vector $U_i(x_j, t)$ represents the displacement of the center of inertia of each elementary cell (pair of atoms), while the vector $u_i(x_j, t)$ represents the mutual displacement for pair of atoms within the elementary cell of relative displacement of sublattices. Hereafter the following standard notations are used: repeated indices assume summation, upper dot stands for the time derivative, spatial derivatives are denoted by means of comma in the indices as it is shown in (7.4.28). The coefficients c_{ijk}, $\bar{c}_{ijk}$, κ_{ikjm}, λ_{ikjm} are the components of the tensors describing elastic properties of the lattice. They possess a certain symmetry under permutation of indices [4]. The nonlinearities in the system are specified by the scalar energy function, $\Phi(u_i)$, describing the interaction of atoms in an elementary cell. It also reflects internal translational symmetry in a complex lattice – the relative displacement of sublattices for a period or for an integer number of periods does not imply change of the complex lattice structure. In a more general case the function $\Phi(u_i)$ may be replaced by another vector periodic function of the argument u. Note that using approximation $\Phi \approx u_i u_i$, Eqs. (7.4.28), (7.4.29) are transformed into a continuum analog of the well-known linear Carman–Born–Kun Huang model [86]. In the case of media with a center of symmetry the second spatial derivatives appear instead of the first ones in both Aero equations [4].

The system (7.4.28), (7.4.29) possesses the conservation law with the following energy integral:

$$E = \int_\Omega \Big(\frac{1}{2} \big(\rho \dot{U}_i \dot{U}_i + \mu \dot{u}_i \dot{u}_i + \lambda_{ikjm} U_{k,j} U_{i,m}$$
$$+ \kappa_{ikjm} u_{k,j} u_{i,m} \big)$$
$$+ c_{ikj} u_k U_{i,j} + \bar{c}_{ikj} u_k u_{i,j} + \Phi \Big) d^3 x. \qquad (7.4.30)$$

In (7.4.30) the symmetries under permutation of indices for material tensors are taken into account.

The existence of a single integral of motion is not sufficient for describing the general solutions behavior when $t \to \infty$. However, it becomes possible after introducing dissipative terms in Eqs. (7.4.28), (7.4.29) proportional to the first time derivatives. Since the system energy is bounded from below for small values of the coupling c between modes, it is possible to justify existence and uniqueness of the system solutions for all $t > 0$ and their convergence to the stationary solutions of Eqs. (7.4.28), (7.4.29).

Below only the most interesting and physically natural case $\Phi' = -a\sin\phi$ with $a > 0$ will be studied. Assume for simplicity that only one component of u and only one component (x or y) of U are distinct from zero and all the solutions depend on only one coordinate $x = x_1$ and on time t.

After adding dissipative terms containing first time derivatives Eqs. (7.4.28), (7.4.29) take the following form:

$$\rho \ddot{U} + \rho_1 \dot{U} = cu_{,x} + \lambda U_{,xx},$$
$$u_{,x} = \partial u/\partial x, \quad U_{,xx} = \partial^2 U/\partial x^2 \qquad (7.4.31)$$
$$\mu \ddot{u} + \mu_1 \dot{u} = -a\sin u - cU_{,x} + \kappa u_{,xx}. \qquad (7.4.32)$$

If $\rho_1 = \mu_1 = 0$ then the system (7.4.31), (7.4.32) possesses the conservation law with the following energy integral:

$$E = \int_\Omega \Big(\frac{1}{2}(\rho\dot{U}^2 + \mu\dot{u}^2 + \lambda U_{,x}^2 + \kappa u_{,x}^2)$$
$$+ cUu_{,x} + a(1 - \cos u)\Big) d^3x. \qquad (7.4.33)$$

If $c = 0$ then the system split into two independent equations: the equation for optical mode becomes a nonlinear wave equation, while dynamics of acoustical mode are described by a linear equation.

Consider the initial-boundary problem for Eqs. (7.4.31), (7.4.32) on the bounded interval $0 \le x \le h$ with the initial conditions

$$U(\chi, 0) = U^0(\chi), \quad u(\chi, 0) = u^0(\chi),$$
$$\partial U/\partial \tau(\chi, 0) = V(\chi), \qquad (7.4.34)$$
$$\partial u/\partial \tau(\chi, 0) = \nu(\chi), \quad \chi = x/h, \qquad (7.4.35)$$

and the boundary conditions

$$(A_1 U + B_1 \partial U/\partial \chi)(0, t) = F_1(t),$$
$$(a_1 U + b_1 \partial u/\partial \chi)(0, t) = f_1(t),$$
$$(A_2 U + B_2 \partial U/\partial \chi)(0, t) = F_2(t),$$
$$(a_2 U + b_2 \partial u/\partial \chi)(0, t) = f_2(t), \qquad (7.4.36)$$

In Fig. 7.4.11 the numerical solution of the system (7.4.31), (7.4.32) is shown for the case of triangular initial distribution of the acoustic displacements $U(\chi)$ (dashed line), zero initial optical displacements $u(\chi)$ (dashed line)

7.4 Control of oscillations in a complex crystalline lattice   151

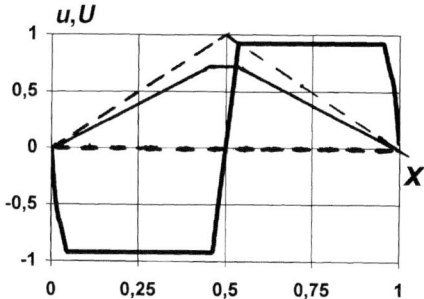

Fig. 7.4.11. Free evolution of optical and acoustic modes.

and zero initial velocities of both variables. Zero values of both functions are also given in the ends of the interval $(0, 1)$. In fact, in the initial time instant, $t = 0$, the deformations $(2\varepsilon = U_{,x})$ have opposite signs in the left and right parts (domains) of the interval. A sharp boundary between domains is interpreted as a defect. The evolution of the deformations for several values of the finite time, $t = T_1$ is shown with solid lines.

Simulations demonstrate strong interaction of the oscillation modes. An inherent structure $u(\chi)$ emerges in the form of two domains of opposite signs (bold solid line). It is interpreted as emergence of the two phases with different values of the order parameter $+u$ and $-u$. Since a strong dependence of the system motion on initial conditions and, certainly, on the coefficients of the equations is observed. In order to eliminate dependence on initial conditions, control theory will be applied.

7.4.2 Control law design

Application of the control methods allows not only the elimination of dependence of solutions on initial conditions, but also it allows the creation of purposeful energy exchange between modes leading to the rebuilding of the lattice structure and to phase transitions. The control goal may correspond to either excitation or to suppression of specific oscillation modes. Let us study the possibility of the optical mode excitation by means of changing the torque applied to the ends of the rod (undimensional lattice) according to a feedback mechanism. To design appropriate feedback the speed-gradient method, Section 2.4.2, will be employed.

Suppose that the bending torque applied to the ends of the rod is considered as the control action. It is assumed that the torque applied to the left end is balanced by the torque applied to the right end, i.e., compression force is identically zero. Then the boundary conditions (7.4.36) read $F_1 = F_2 = 0$, $f_1(t) = -f_2(t) = f$, where f is the control variable. Let the control goal be formulated as an increase of the optical mode energy E. According to the speed-gradient method the goal function Q should be introduced such that the achievement of the control goal corresponds to maximization of the goal function. In this case the total energy can be chosen as the goal function, i.e., $Q = E$, where E is defined in (7.4.30).

$$K = (hs/2) \sum_{i=1}^{n} \dot{u}_i^2. \tag{7.4.37}$$

According to the speed-gradient method, the asymptotic maximization of the functional (7.4.37) can be achieved by choice of the controlling action's sign which coincides with the sign of the speed-gradient (gradient of the speed of changing Q along the solutions of the system)

$$f = -R\left(\nabla f \dot{Q}(x, f)\right) \tag{7.4.38}$$

where x is the state vector (function) of the controlled system (7.4.31), (7.4.32), f is the vector of controlling variables, R is a vector-function forming an acute angle with its argument, e.g., multiplying by a positive factor or taking sign of each component of the vector argument.

For the sake of simplicity the problem is discretized by means of replacing the first and second derivatives in the system equations (7.4.31), (7.4.32) and in the expression for the energy (7.4.30) by their finite differences as follows: $(u_{i+1} - u_i)/h$; $(u_{i+1} - u_i)/T$; $(u_{i+1} - 2u_i + u_{i-1})/h^2$; $(u_{i+1} - 2u_i + u_{i-1})/T^2$.

Direct calculation yields the following form of the speed-gradient function:

$$\psi = (1/h)(\dot{u}_1/a_1 - \dot{u}_n/a_n) \tag{7.4.39}$$

Therefore, the control algorithm for excitation of the optical mode may have for example, the "relay" form

$$f = \gamma \operatorname{sign}(\dot{u}_1/a_1 - \dot{u}_n/a_n) \tag{7.4.40}$$

Numerical results are obtained for the excitation of oscillations in the discrete version of the system consisting of $n = 50$ atoms. The following constants in the equations are chosen: $\rho = 0.9$, $\rho_1 = 0$, $\mu = 2.5$, $\mu_1 = 1$, $c = 5$, $a = 1.5$, $\lambda = \kappa = 1$, $A_1 = A_2 = 1$, $a_1 = a_2 = 1$, $B_1 = B_2 = b_1 = b_2 = 0$.

The control algorithm $f = 0.5 \operatorname{sign}(\dot{u}_1/a_1 - \dot{u}_n/a_2)$ is used. In Fig. 7.4.12, the shapes of the acoustic mode $U(\chi)$ and the optical mode $u(\chi)$ are shown for time $t = 15$. In the initial time $t = 0$ both functions are equal to zero. The evolution of the energy of the modes is shown in Fig. 7.4.13. The effect of the control leads to a strong excitation of the optical mode.

An important question in crystalline lattice dynamics is sensitivity to changes of initial conditions. A number of simulations have been performed to examine dependence of the limit energy of modes on the initial conditions. Figure 7.4.14 shows the results for two different triangular functions $U(\chi)$, like in Fig. 7.4.11. At the initial time $t = 0$ zero optical displacements $u(\chi)$, $U(\chi)|_{\chi=0.5} = 1$ (line a) and $U(\chi)|_{\chi=0.5} = 10$ (line b) are taken. Initial velocities of both variables are zeroed. It is seen that changes of initial displacement in order of magnitude leads in $3 \div 5\%$ changes of limit energy of each mode.

7.4 Control of oscillations in a complex crystalline lattice 153

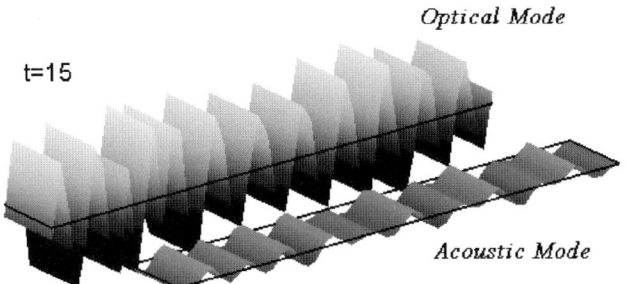

Fig. 7.4.12. Controlled excitation of optical mode.

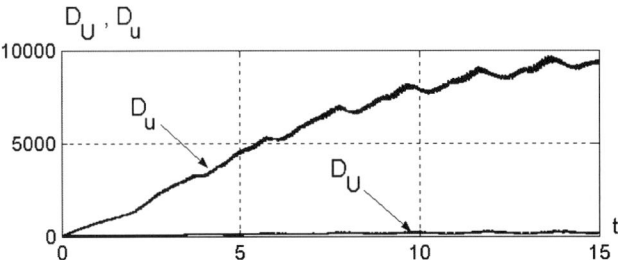

Fig. 7.4.13. Evolution of energy of the modes in the controlled system.

7.4.3 Nonfeedback control

For the study of microscopic systems the problem of physical realization of control arises. The main difficulty is to implement the feedback exploiting measurements of the microscopic phase variable deflections. To solve analogous problems in the area of molecular and quantum control [53, 91, 147] the idea of using a program (feedforward, nonfeedback) control was proposed. In our case the idea is to first calculate controlling action $f(t)$ as a function of time during the simulation of the system with the feedback algorithm (7.4.40). Then the precalculated function $f(t)$ is applied to the physical system during the experiment. At the second stage neither measurements nor feedback is used.

To analyze the efficiency of such an approach the simulation of the system with a nonfeedback control action was performed. The results are shown in Fig. 7.4.15. Initial conditions were chosen to be the same as for feedback control, both for calculation of program control $f(t)$ and for its testing. It is seen that nonfeedback control designed using the proposed method provides qualitatively the same results as with feedback algorithm (7.4.40) (cf. Fig. 7.4.14).

Conclusions. The possibility of purposeful excitation of the optical mode by means of torque applied to the ends of the lattice (rod) in a broad range

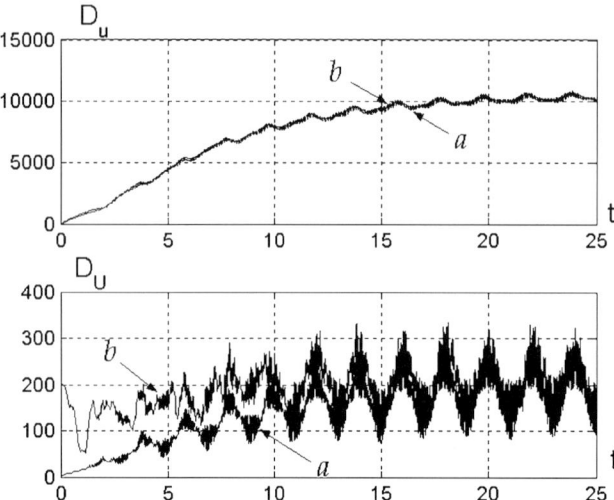

Fig. 7.4.14. Dependence of energy of the modes on the initial conditions in the controlled system.

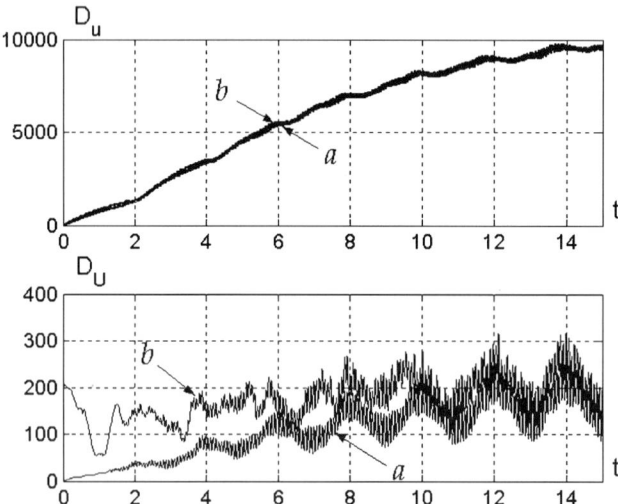

Fig. 7.4.15. Dependence of energy of the modes on the initial conditions in the nonfeedback system.

of initial conditions on the steady-state oscillations has been demonstrated. It means that application of control may allow the elimination or reduction of the influence of initial conditions. A nonfeedback control algorithm is described possessing similar properties.

Another consequence stems from the well-known fact that static strains (deformations) influence the phase state of smart materials. It implies that application of control of energy exchange between macroscopic deformation and microscopic degrees of freedom allows the control of dynamics of phase transitions and, particularly, the effect of the memory shape which is characteristic for smart materials.

7.5 Control of chaos in distributed systems

7.5.1 Spatiotemporal systems

Methods of early works on spatiotemporal control of chaos were similar to the finite-dimensional case: OGY/OPF, delayed feedback and others, see survey [200]. In the consequent papers other approaches were introduced and investigated (mainly numerically).

In the paper [338] a one-dimensional array of $N = 100$ cells, described by logistic maps $(F(x, u) = 1\alpha x^2 + u)$, where the value of the parameter α ensured chaotic behavior of each cell for $u \equiv 0$ was considered. It was shown by numerical experiments that local feedback:

$$u_i(n) = \gamma \left[x_i(n) - \frac{1}{N+1} \sum_{j=0}^{N} x_j(n-1) \right] \quad i = 1, 2, \ldots, N-1 \quad (7.5.41)$$

provides stability of the spatially uniform distribution $x_i \equiv x_*$, $i = 0, 1, 2, \ldots, N$ for sufficiently large gain $\gamma > \gamma_0$. For $\gamma < \gamma_0$ a nonuniform distribution consisting of several clusters of uniformity is stabilized, while each cell is still periodically oscillating. Similar behavior has been observed with local feedback in error:

$$u_i(n) = \gamma \left[x_i(n) - x_* \right], \quad (7.5.42)$$

or with the so-called *global* feedback depending on observable average values of variables:

$$u_i(n) = -\frac{\gamma}{N+1} \sum_{j=0}^{N} \left[x_j(n) - x_j(n-1) \right], \quad (7.5.43)$$

or

$$u_i(n) = -\gamma \left[\frac{1}{N+1} \sum_{j=0}^{N} x_j(n) - x_* \right]. \quad (7.5.44)$$

The latter ones look more realistic. The above results were justified analytically in [172]. Cluster synchronization for the chains of chaotic oscillators was also reported in [328, 329].

In 1997 L. Kocarev et al. [238] studied pinning control for one-dimensional lattice of Lorenz systems, for control interacting with every p-th cell. Attractivity of the spatially uniform (coherent) yet chaotic in time motion under discrete-time control (7.5.42) with $\gamma = 1$ applied to the first equation of the Lorenz system was established. Similar result for two-dimensional lattice of Lorenz systems was obtained in [412] using an integral feedback (called "adaptive" by the authors). Analogous results were established for the complex Ginzburg–Landau equation:

$$\dot A = A + (1 + i\mu_1)\frac{\partial^2 A}{\partial r^2} - (1 + i\mu_2)|A|^2 A, \qquad (7.5.45)$$

see [303] and for the Swift–Hohenberg equation describing dynamics of some types of semiconductor lasers [70]. Complex Ginzburg–Landau equation may describe a number of phenomena in laser physics, hydrodynamics, chemical turbulence. It can exhibit different forms of complex behavior, including Andronov–Hopf bifurcation, chaotic turbulent modes, contrast structures. Complex Ginzburg–Landau equation with pinning control, applied only in a finite number of points was studied in [79, 450]. The maximum distance between nodes of control ensuring achievement of the control goal was found numerically. A similar result for boundary control was obtained later in [450].

A possibility of stabilization of solutions to the Kuramoto–Sivashinsky equation

$$\frac{\partial \varphi}{\partial t} + \varphi\frac{\partial \varphi}{\partial r} + \frac{\partial^2 \varphi}{\partial r^2} + \frac{\partial^4 \varphi}{\partial r^4} = u \qquad (7.5.46)$$

by periodic-delayed velocity feedback

$$u = \varepsilon^t \frac{\partial \varphi}{\partial t}(t - \tau), \qquad (7.5.47)$$

where τ is time delay was demonstrated in [393].

Pinning controls (local injections) were also applied in the paper [201] to stabilization of the trivial solution ($x_i(t) \equiv 0$) of coupled oscillators systems with diffusion-gradient coupling:

$$\dot x_i = f(x_i) + \frac{\varepsilon}{2}(x_{i-1} - 2x_i + x_{i+1}) + \frac{\rho}{2}(x_{i-1} - x_{i+1}) + u_i, \qquad (7.5.48)$$

as well as to the CGL equation evolving initially in a chaotic mode. Linear high-gain feedback in each lth oscillator was employed. Stability analysis was performed based on linearized models near the goal solution.

Minimal density of the local control nodes and the optimal allocation were determined in [185] for one-dimensional array of coupled logistic systems: $f(x) = ax(1 - x)$ in (7.5.48), using a linear feedback for stabilization. The method of stabilization of space-homogenious solution of the reaction–diffusion equation was proposed in [283] for the complex Kuramoto–Suzuki equation. In [443] the method of chaos and spiral waves suppression by a weak distributed perturbation for Maxwell–Bloch equation with difraction coupling was proposed.

Some interesting control problems arise when studying *cluster synchronization* of two- or three-dimensional arrays of nonlinear oscillators. For example, in [63–65, 352] the conditions, guaranteeing splitting the array into a given number of synchronously oscillating clusters are proposed. Based on Lyapunov functions of special form it is shown that growth of the coupling strength k leads to an increase of the number of the clusters, up to complete synchronization. Though the authors of the aforementioned papers do not mention control explicitly, their results can be interpreted as design of the coupling strength k to ensure the given number of clusters. Future studies could be devoted to the design of adaptive control of clustering by means of changing the coupling strength online, during the experiments with the system.

Control can promote or suppress even more complex spatiotemporal behavior. Intricate patterns of wave propagation in a chemical reaction–diffusion system with spatiotemporal feedback were found in [386]. Wave behavior is controlled by feedback-regulated excitability gradients that guide propagation in specified directions. In [68, 232] it was demonstrated that chemical turbulence can be completely or partially suppressed by the feedback, giving rise to stable uniform oscillations, intermittent regimes with reproduction cascades of amplitude defects, or regular patterns of clusters and standing waves. Similar effects are expected for other reaction–diffusion systems of different origins. Moreover, experiments show possibility of control for other sophisticated types of behavior, e.g., control of polarization switching in vertical cavity surface emitting lasers [285, 337].

7.5.2 Delayed systems

Delayed (or more generally retarded) systems are described by infinite-dimensional models, typically by delay-differential systems, e.g.,

$$\dot{x} = F(x(t), x(t-T), u(t), u(t-T)), \tag{7.5.49}$$

where T is a time delay. Model (7.5.49) can be reduced either to discrete finite-dimensional system of type (7.5.49) if delay time is a multiple of the sampling interval. Otherwise, it can be converted into a spatiotemporal model by transformation $t = \sigma + \theta T$, where $0 \leq \sigma \leq T$ is a continuous space-like variable, and θ is a discrete temporal variable. An example of delayed models is CO_2 laser.

Delay may also appear due to introducing a delayed feedback (see Section 6.4.3). A typical example of a chaotic delayed system is the Mackey–Glass system:

$$\dot{y} = -\gamma y + \beta y(t-T)/(\alpha + y(t-T)^n)),$$

with positive parameters $\alpha, \beta, \gamma, n, T$. This system was studied by C. Mackey and L. Glass [281] in 1977 as a model of biological chaotic processes.

In papers [78] two version of adaptive control procedure have been proposed based on results of perturbing the value of delay time for the Bénard–Marangoni time-dependent convection. Both procedures are based on results of [36]. An experiment showed that both procedures ensure suppression of phase defects and stabilization of regular oscillations. M. Ciofini et al. [109]

7 Control of Interconnected and Distributed Systems

proposed a control method for a delayed dynamical system exhibiting high-dimensional chaos. The control is based on a negative feedback loop with adaptive filtering consisting of a selective filter centered at the frequency of the orbit to be stabilized with the addition of a time-derivative correction.

7.5.3 Chaotic mixing

An important practical field of research is chaotic mixing, particularly the mixing of fluids and granular flows. Mixing properties of flows are important in a variety of applications, such as chemical production in continuously stirred chemical reactors, production of powders, polymers, design of combustion processes, and heat exchangers. A typical control goal is to increase the rate and quality of mixing. Thorough mixing is of utmost importance in chemical engineering; it is recognized that 80% of the costs of chemical products falls into their purification. After quality mixing the amount of nonreacted species in the product decreases and, therefore, the impurity of the product decreases too.

In 1997, A. Sharma and N. Gupte [399] proposed a method of enhancing the rate of mixing of chaotic flows by increasing their chaoticity. Chaoticity is measured by local Lyapunov exponents which estimate the average rate of stretching. Using the Lagrangian description of two-dimensional fluid dynamics, fluid particle trajectories are described by the corresponding two-dimensional Hamiltonian equations. In [399] a generalized description of mixing by a nonlinear dynamical system is considered:

$$\dot{x} = F(x(t), u(t)), \tag{7.5.50}$$

where $x(t) \in \mathbb{R}^n$ is state vector, $u(t)$ is controlling parameter of the flow, together with its variational system

$$\dot{w} = M(x, u)w, \tag{7.5.51}$$

where $M = \partial F/\partial x(x, u)$ is Jacobian. Sharma and Gupte suggested to measure the local stretching rate by increase rate of the w squared norm: $\frac{\partial}{\partial t}|w|^2 = 2w^\mathrm{T} M(x, u)w$ and to change the parameter according to the rule

$$\Delta u = \gamma \mathrm{sign}(w^\mathrm{T} \frac{\partial M}{\partial u} w), \tag{7.5.52}$$

where $\gamma > 0$. Control is activated in the regions where the largest local Lyapunov exponent is less then its average value. It is easy to see that algorithm (7.5.52) is a special case of the speed-gradient algorithm (see Section 2.4.2), the goal function to be maximized being $Q(x) = |x|^2$. In [399] a discrete-time control algorithm that turns out to be a special case of gradient algorithm (see Section 2.4.1) is also described. It is applied to enhance chaoticity of the Chirikov (*standard*) map. A disadvantage of this approach is that in both

7.5 Control of chaos in distributed systems

continuous-time and discrete-time versions the state vector of the system is supposed to be available for measurement and the system parameters should be known, which is difficult to provide in practice.

Mixing of granular materials provides fascinating examples of pattern formation and self-organization. For example, more mixing action (increasing the forcing with more vigorous shaking or faster tumbling) does not guarantee a better-mixed system. This is because granular mixtures of different materials segregate according to density and size and, in fact, the very same forcing used to mix may lead to unmixing. Self-organization results from two competing effects: chaotic advection or chaotic mixing, as in the case of fluids, and flow-induced segregation. The rich array of behaviors is ideally suited for nonlinear dynamics-based inspection. In fact, these systems may constitute the simplest example of coexistence between chaos and self-organization that can be studied in the laboratory.

The dependence of mixing quality on various parameters has been studied (numerically or experimentally) by many authors. It has been demonstrated that the flow in elliptical and square mixers is time periodic and results in chaotic advection and rapid mixing unlike the flow in circular mixers. As for control by external forcing, most authors consider the case of open-loop (feedforward) periodic control. Perhaps the most elaborated treatment of the control problem for optimal mixing was performed in [115] where a prototypical mixing problem in an optimal control framework was formulated. The objective in [115] is to determine the sequence of fluid flows maximizing the entropy. By developing appropriate ergodic-theoretic tools for the determination of entropy of periodic sequences, the authors derive the form of the protocol maximizing entropy among all of the possible periodic sequences composed of two shear flows orthogonal to each other.

8

Control of Molecular and Quantum Systems

8.1 Laser control of molecular dynamics

The interest of humanity in control of microworld processes has a rich history. We already mentioned briefly the history of Maxwell Demon starting from the end of the 19th century. In the 20th century numerous control problems in chemical reactors and nuclear reactors have been studied. Control goals in conventional applications are usually formulated as regulation of process intensity in cases when the normal functioning of the process is in principle possible even without control. Challenging problems, however, are to change the natural course of the process, to intervene the motion of single atoms and molecules, to break existing chemical bonds and to create new ones [91, 114, 367]. In the 20th century implementation of these goals has become a matter of serious discussion owing to the invention of such fine instruments as lasers.

Major difficulties when controlling processes at the atomic or molecular level are caused by the tiny spatial size of the controlled objects and the fast speed of the processes in the microworld. Indeed, an average size of a molecule of a chemical substance (monomer) is of the order of 10^{-8} m $=$ 10 nm. An average interatomic distance in the molecule is of order 1 nm, an average speed of atoms and molecules at the room temperature is about 10^2–10^3 m/s, and an average period of natural oscillation of a molecule is 10–100 fs (1 fs $= 10^{-15}$ s). Development of devices for measurement and control at such spatiotemporal scales is an enormously hard scientific and technological problem.

The situation changed in the end of the 1980s with invention of ultrafast femtosecond lasers generating pulses of the order of tens or now even units of femtoseconds as well as the methods of computerized control of the laser pulses shape. Control is performed by changing the shape of laser pulses by means of liquid crystal or acousto-optical computer-controlled modulator. A new avenue of chemical research, the so called *femtochemistry*, arose. A. Zewail was awarded the 1999 Nobel Prize for the progress in this area [460]. The advances of femtosecond lasers applications gave rise to the term "femtosecond technologies" or "femtotechnologies". These technologies are used, in particular, to solve the problems of selective dissociation where one needs to break certain molecular bonds without affecting, wherever possible, the rest.

Several approaches to the control of molecular systems were suggested. M. Shapiro and P. Brumer [398] used control on the basis of interference of two laser beams of different frequencies, amplitudes, and phases (the pumping-damping scheme). D.J. Tannor and S.A. Rice [425] suggested two-pulse schemes of pumping-damping in the time domain. Methods of optimal control, based, in particular, on the V.F. Krotov method [241, 243, 244] were later used in pulse optimization. H. Rabitz and his collaborators [114, 216, 344, 366] considered various versions of optimal control under the classical and quantum descriptions of the dynamics of molecular motion. H. Rabitz put forward [216] the idea of realizing the adaptive laser control of chemical reactions by means of the methods of search optimization (genetic algorithms). Feasibility of the approach was corroborated experimentally [38, 54, 342].

Since the beginning of the 1990s there has been a growing interest in the control problems for molecular systems in classical and quantum formulation [114, 262, 366, 372, 455]. One of the benchmark problems in the field is the dissociation problem for diatomic molecules [180, 181, 188, 242, 274, 455]. The possibilities of the dissociation of a molecule by monochromatic (single frequency) laser field have been explored for the case of a hydrogen fluoride (HF) molecule using Chirikov's resonance overlap criterion in the paper [180]. The case of a two-frequency (two-laser) control field was investigated in [181, 188]. It was shown that the intensity of a bichromatic field required for dissociation can be reduced compared to the monochromatic case. In [274] the possibility of further reduction of the control field intensity by means of chirping (frequency modulation) of the laser frequency with constant chirping rate has been demonstrated.

New possibilities for the changing of physical and chemical properties are provided by using feedback for control design. In [114, 455] methods based on geometric control theory (inverse control) were proposed for control of molecular systems, including dissociation problems for diatomic and triatomic molecules. In a number of papers the possibility of optimal control design is discussed [241, 344]. Two methods for dissociation of diatomic molecules design by feedback control based on resonance curve and speed-gradient principle were proposed in [154, 155]. The common feature of the methods lies in that they are used to design the control action as a time function from the given model of the molecular system. In the computer experiments, one may assume that all the necessary signals are measured and the design algorithm is realized by the computer. As a result, the control signal will be generated as a time function. At the second stage the implementation of control as applied to the real system is done without measurements and feedback. Numerous uncertainties hinder the practical application of the methods: the initial system state is not known precisely; the constructed control function is calculated inaccurately and realized with error; the model of the molecule itself is imprecise because its parameters are not known precisely. Even the very choice between the classical and quantum descriptions is the matter of repeated discussions.

First results on control of quantum-mechanical systems were published in the late 1970s [46, 94] and in the 1980s [95, 202, 345]. Recently, interest in the control of quantum systems was growing rapidly. According to the Science Citation Index, by the beginning of this century more than 500 papers were published annually in peer-reviewed journals. In addition to applications in chemical technologies, studies related to the quantum computers motivate the development of the field [234]. The state-of-the-art is presented in books and collections of papers [53, 92, 360]. Some open problems are listed in the surveys [91, 280, 368]. There is also a number of recent experimental works on quantum control devoted, e.g., to control of coupled spin dynamics in NMR [280], to manipulating the quantum state of trapped ions via lasers and electric fields [370], to open-loop control in superconducting circuits [441].

Instead of trying to make a survey of an avalanche of publications, we describe below a new approach to control of molecular and quantum systems based on speed-gradient method with energy-related goal functions. Such a version of speed-gradient method has already been used several times in this book. The proposed algorithms feature robustness because they are independent of the shape of the potential of intermolecular interaction. They enable one to achieve dissociation with a smaller intensity of the control field than in the case of chirping and, as compared with the methods of optimal control, are easier to design and calculate.

First, we describe the algorithms based on classical models of molecular dynamics. Classical models are often used instead of the quantum ones in dynamics calculations of molecular motion [180, 181, 395, 430, 440]. It has been shown previously [180, 181, 440] that the results provided by classical and quantum models are close for model systems with one or two degrees of freedom. Even when the expectation values of the quantum wave packet do not exactly correspond to the averages over classical trajectories, observables such as dissociation probabilities are shown to be quite similar for monochromatic [180, 440] and bichromatic [181] excitation. In the next section we further investigate the possibility of using classical models for control algorithm design. To this end the comparison of the results obtained by simulation of classical and quantum-mechanical ensembles is performed.

In the final section of the chapter we describe an alternative control design based on quantum-mechanical system description. Again it turns that the speed-gradient method can be used to design a feedback control law generating an open loop control action as a function of time.

8.2 Controlled dissociation of diatomic molecules (classical design)

8.2.1 Control algorithm design

The first step of the approach is to reformulate the controlled dissociation problem as the one of achieving the given level of the molecular energy (dissociation threshold). To simplify study assume that the given energy level is slightly less than the dissociation threshold, i.e., the *predissociation* problem is considered. The goal function is chosen as the squared deflection of the current energy from the desired one. Then the standard speed-gradient control algorithm, see Section 2.4.2 is designed and the control system with the "reference" molecule is simulated during some time T_1, sufficient for its dissociation. Being applied to a real molecular system such an action will cause dissociation of only a fraction of all molecules, namely, those with initial states close to the initial state x_0 of the reference molecule. Now assume that control is implemented as a series of repeating pulses of length T_1 and the intervals between pulses are large enough. During the pauses between pulses the states of some portion of molecules will approach the point x_0 in the state space in the course of their chaotic thermal motion. Then the next pulse after the pause will lead to dissociation of that portion of molecules. Such behavior of the controlled system will take place if the closed-loop system is sufficiently rough (robust). Robustness of the molecular system depends on the shape of the pulse. If the system is robust enough, then each pulse will cause dissociation of a substantial portion of molecules and a proper quality of control will be achieved. Let us consider this approach in more detail, following [18, 126].

Controlled system model. Let us start with classical description of a diatomic molecular system under the action of the external laser field. Dynamics of such a system can be described by the following controlled Hamiltonian [188, 455]

$$H = \frac{p^2}{2m} + \Pi(r) - \mu(r)\, u(t), \qquad (8.2.1)$$

where the coordinate $r(t)$ is the interatomic distance, p is momentum, $\Pi(r)$ is potential of interatomic interaction, m is the mass of the molecule, $\mu(r)$ is dipole moment of the molecule, $u(t)$ is intensity of external field. The value $u(t)$ serves as control variable. Substitution of (8.2.1) into the Hamiltonian equations

$$\frac{\partial r}{\partial t} = \frac{\partial H}{\partial p}, \quad \frac{\partial p}{\partial t} = -\frac{\partial H}{\partial r} \qquad (8.2.2)$$

yields the following equation of molecular motion

$$m\ddot{r} = -\Pi'(r) + \mu'(r)u(t). \qquad (8.2.3)$$

For description of intermolecular interaction we use the standard Morse interatomic potential model

8.2 Controlled dissociation of diatomic molecules (classical design)

$$\Pi(r) = D\left(1 - e^{-\alpha(r-a)}\right)^2 - D = D\left(e^{-2\alpha(r-a)} - 2e^{-\alpha(r-a)}\right), \qquad (8.2.4)$$

where D is the bond energy, a is the equilibrium interatomic distance. For the dipole moment a linear approximation is often used [188, 455]

$$\mu(r) = Are^{-\xi r^4} \quad \mu'(r) = A\left(1 - 4\xi r^4\right)e^{-\xi r^4}, \qquad (8.2.5)$$

where A, ξ are constant parameters. Thus the equation of motion in the Lagrange form reads

$$m\ddot{r} = 2\alpha D\left(e^{-2\alpha(r-a)} - e^{-\alpha(r-a)}\right) + Au(t). \qquad (8.2.6)$$

Adopting such a description we consider only one-dimensional motion of the molecules, and assume that the symmetry axis of the molecule is oriented along the force lines of the controlling external field. It means that we neglect the effects of the rotation and changing orientation of the molecules.

Control goal. In the case of the dissociation problem the natural goal of control can be expressed in terms of the molecule free energy. Note that as the molecule energy approaches the level $\Pi_* = \lim_{r\to\infty} \Pi(r)$, the dissociation is getting more probable. In the case of the Morse potential (8.2.4), obviously $\Pi_* = 0$.

Let us choose the goal function as the squared error in terms of energy: $Q(q,p) = 0.5(H_0(q,p) - H_*)^2$, where

$$H_0(q,p) = \frac{p^2}{2m} + \Pi(r)$$

is the total energy of the free molecule, H_* is the prespecified value, slightly less than the dissociation threshold Π_*.

Control algorithms. Evaluating the rate of change of the goal function for constant u and then the speed-gradient as previously we arrive at simple feedback control laws

$$u = -E\left(H_0(q,p) - H_*\right)\dot{r}, \qquad (8.2.7)$$

$$u = -E\,\text{sign}\left(H_0(q,p) - H_*\right)\,\text{sign}\,\dot{r}, \qquad (8.2.8)$$

where $E > 0$; $\text{sign}(H) = 1$ for $H > 0$, $\text{sign}(H) = -1$ for $H < 0$ and $\text{sign}(0) = 0$.

In what follows we use a simplified version of the algorithm (8.2.8) obtained under following assumption: the current energy of the molecule is always less than its goal value H_*:

$$u = E\,\text{sign}\,\dot{r}. \qquad (8.2.9)$$

The algorithm (8.2.9) does not require exact knowledge of the goal value H_* and acts as introduction of a negative Coulomb friction into the system. The method can be applied to other problems, e.g., localization of the molecule in the region of higher energy.

8.2.2 Simulation results (classical model)

A series of computer experiments for the system (8.2.6), (8.2.9) were performed. Parameter values were chosen corresponding to the hydrogen fluoride (HF) molecule [188, 455]: $m = 1732$, $D = 0.2101$, $\alpha = 1.22$, $a = 1.75$, $A = 0.4541$, $\xi = 0.0064$, $E = 0.1$. The values are indicated in Hartree atomic units (a.u.). Initial values of the "reference molecule" for calculation of the controlling function $u(t), 0 \leq t \leq T_1$ were taken near equilibrium state $r = a$, $\dot{r} = 0$. The intensity of the field was chosen sufficiently small: $E = 0.005$ a.u.

The designed controlling function $u(t)$ was applied to the ensemble of the molecules consisting of $N = 1000$ molecules. The interaction between molecules and interaction between molecules and the boundary were neglected. Initial conditions for the ensemble were chosen randomly and distributed uniformly over the given energy surface $H_0 = -0.8689D$. Control was applied as a series of repeating pulses with the period T_2. The value of T_2 should be chosen large enough in order to allow the molecules to mix during pauses between pulses. A typical value was $T_2 = 200\, T_0$, where T_0 is the period of small oscillations of the molecule near equilibrium.

The efficiency of control was measured as the ratio of the number of dissociated molecules over the total number of molecules (in percent). A molecule was considered as dissociated[1] if its energy exceeded the level $H_* = -0.1185D$.

The efficiency of the proposed algorithm (8.2.9) was compared with the efficiency of the standard chirping algorithm

$$u(t) = E \cos(\phi_0 + \Omega_0 t - \frac{\varepsilon t^2}{2}) \tag{8.2.10}$$

The time dependence of the portion of the dissociated molecules when controlling with a linearly chirped field is shown in the Fig. 8.2.1(a). The chirp rate ε was tuned to achieve maximum portion of dissociated molecules. The final choice was $\varepsilon = 0.01 \Omega_0 / T_0$. Similar time dependence for the speed-gradient algorithm (8.2.9) is shown in the Fig. 8.2.1(b). It is seen from the figures that the speed-gradient algorithm provides several times higher efficiency than the chirping algorithm. It is important to notice that chirping control is very sensitive to the changes of the chirping rate ε. The choice of the value of ε requires time-consuming computations and more precise knowledge of the molecular Hamiltonian and dipole moment than it is needed for efficient work of the algorithm (8.2.9).

8.2.3 Comparison of classical and quantum simulations

An interesting and still disputable issue is the possibility of using classical modeling of molecular processes instead of quantum-mechanical ones. In the

---
[1] Since the value H_* is less than the dissociation threshold $H_* = 0$, such a state can be termed *predissociation*. However, we will not discuss this difference further, because the goal of the research is just evaluation of the feasibility of the approach.

8.2 Controlled dissociation of diatomic molecules (classical design) 167

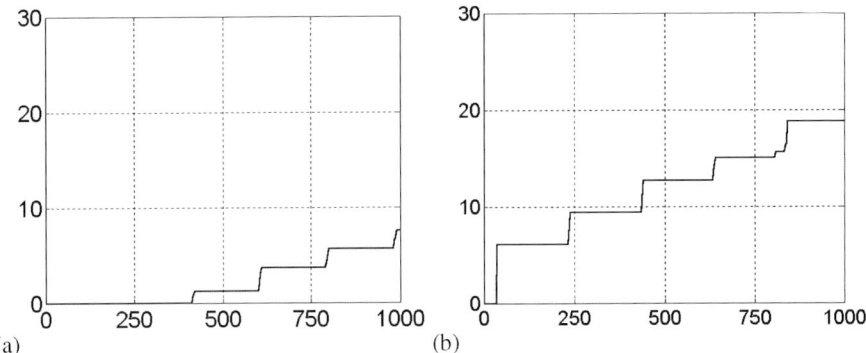

Fig. 8.2.1. Controlled dissociation of the classical ensemble (a) – linearly chirped pulses, (b) – pulses calculated by the speed-gradient method. The portion of the dissociated molecules is shown in percent, time is shown in the units of T_0.

case of a diatomic molecule described classically by the model (8.2.6) it is known that its dynamics can be described more adequately by the quantum-mechanical (more precisely, semiclassical) model represented as the time-varying (controlled) Schrödinger equation

$$i\hbar \frac{\partial \Psi}{\partial t} = \frac{\hbar^2}{2M} \frac{\partial^2 \Psi}{\partial r^2} + \Pi(r)\Psi + Aru(t)\Psi, \qquad (8.2.11)$$

where $\Psi = \Psi(t,r)$ is the wave function, $\Pi(r)$ is the Morse potential (8.2.4). Recall that the squared absolute value of the wave function $|\Psi(t,r)|^2$ defines the probability density of the molecule in the given state. The probability of dissociation is defined as the probability of the molecule to have energy greater than the dissociation threshold H_*.

It turns out, however, that the classical calculations give in many cases the result close to that of the quantum-mechanical ones. Therefore, numerical comparison of the controlled dissociation rates for classically and quantum-mechanically modeled systems has been performed.

To analyze the quantum-mechanical model (8.2.11) a finite-level approximation of the model was developed based on the expansion of the solution to the time-varying Schrödinger equation over the eigenfunctions of the free Schrödinger equation (with $u(t) = 0$). Eigenvalues and eigenfunctions of the free Schrödinger equation with the Morse potential can be evaluated analytically [131]. The controlling function and the final simulation time were chosen the same as in the classical case. The initial system state was chosen as the pure state with the energy equal to the energy of the second energy level, while the dissociation threshold H_* was chosen as the 15th energy level for the HF molecule, which also corresponds to the classical case.

The results of the quantum-mechanical simulation are shown in the Fig. 8.2.2. It is seen that the speed-gradient algorithm provides a dissociation

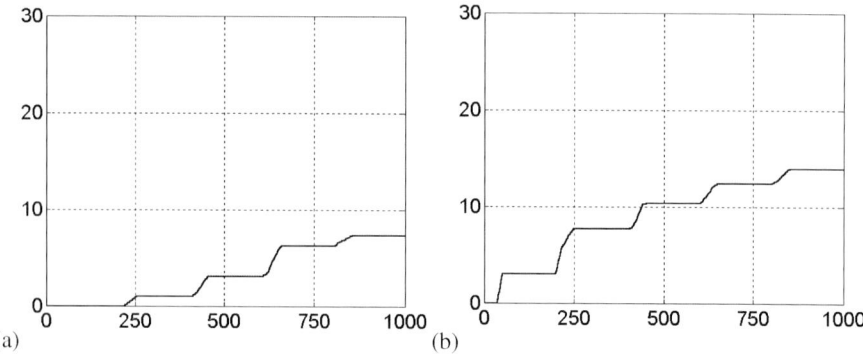

Fig. 8.2.2. Dissociation probability for quantum-mechanical modeling for $E = 0.005$ a.u.: (a) – linearly chirped pulses, (b) – pulses calculated by the speed-gradient method. The portion of the dissociated molecules is shown in percent, time is shown in the units of T_0.

probability of 14% after 5 pulses which exceeds the dissociation probability for the chirped pulse significantly and corresponds to the results for the classical case (10–12%).

8.3 Control of finite-level quantum systems

An approach to control of observables for a class of quantum-mechanical systems and its application to controlled dissociation of diatomic molecules based on quantum-mechanical description are described below. The exposition follows [19, 20].

Let the dynamics of the controlled system satisfy the finite-dimensional controlled Schrödinger equation:

$$i\hbar \dot{\Psi}(t) = [H_0 + \sum_{k=1}^{r} u_k H_k]\Psi(t), \qquad \Psi(t) \in \mathbb{C}^n, \qquad (8.3.12)$$

where $i = \sqrt{-1}$ is imaginary unit, $\hbar$ is Planck constant, H_0 is the Hamiltonian operator (energy operator), defining the dynamics of the free (noncontrolled) system, u_k are real-valued scalar control functions and the terms $u_k H_k$ are interaction Hamiltonians. The operators H_k for $k = 0, \ldots, r$ are self-adjoint. The phase space of this system is represented by the unit sphere of the n-dimensional complex space $\mathbb{C}^n$. The following control problem is posed: to construct functions $u_k(\Psi)$, ensuring that the average of the given observable Z tends to its goal value Z_*. Precisely,

$$\lim_{t \to +\infty} \Psi(t)^* Z \Psi(t) = Z_* \qquad (8.3.13)$$

8.3 Control of finite-level quantum systems

for all initial conditions $\Psi(t_0)$, where Z is a self-adjoint operator. Recall that the term "observable" in quantum mechanics is used to denote an operator in the state space of the system, corresponding to some characteristic of the system: the system energy, velocity, angular momentum, etc. Any measurement results in a specific number – the value of the observable which is a random number. Then the value $\Psi(t)^* Z \Psi(t)$ has the meaning of its average at time t.

To design a control algorithm for the problem (8.3.12), (8.3.13) we use the speed-gradient method, see Section 2.4.2. Introduce the following goal function:

$$Q(\Psi) = (\Psi^* Z \Psi - Z_*)^2. \quad (8.3.14)$$

Then the goal (8.3.13) is reformulated in the form

$$\lim_{t \to +\infty} Q(\Psi(t)) = 0.$$

Assume that the observable Z commutes with the operator H_0, i.e., the goal observable is a function of the system energy. Evaluate the speed $\dot{Q}$ of changing the function (8.3.14) along trajectory of the system (8.3.12) and then the partial derivatives of $\dot{Q}$ with respect to u_k, $k = 1, \ldots, r$:

$$\nabla_u \dot{Q} = -\gamma \nabla_{u_k} \{2(\Psi^* Z \Psi - Z_*)(\dot{\Psi}^* Z \Psi + \Psi^* Z \dot{\Psi})\} =$$

$$-\gamma \nabla_{u_k} \left\{ \frac{2i}{\hbar} (\Psi^* Z \Psi - Z_*) \Psi^* [H_0 Z - Z H_0 + \sum_{k=1}^{r} u_k (H_k Z - Z H_k)] \Psi \right\}.$$

We arrive at the following control algorithm in the finite form:

$$u_k(\Psi) = -\gamma \frac{2i}{\hbar} (\Psi^* Z \Psi - Z_*)(\Psi^* [H_0 Z - Z H_0 + (H_k Z - Z H_k)] \Psi). \quad (8.3.15)$$

Thus, we have designed the feedback algorithm of controlling the average value of observable Z for the quantum system with finite-dimensional state space. Note that the commutator of self-adjoint operators multiplied by $i/\hbar$ can be interpreted as quantum Poisson bracket, i.e. algorithm (8.3.15) is analogous to the energy control algorithm (3.1.10). Algorithm (8.3.15) can be used for control of the systems with infinite-dimensional state space as well.

To formulate conditions ensuring achievement of the control goal, impose the following restrictions on the operators Z, H_0, and H_1:

A1) $z_k - z_m \neq 0$, $k \neq m$, $k, m = \overline{1,n}$ where z_k are eigenvalues of Z in ascending order;

A2) $(\lambda_k - \lambda_m) \neq (\lambda_r - \lambda_s)$, $(k,m) \neq (r,s)$, $k, m, r, s = \overline{1,n}$, where λ_k are eigenvalues of H_0 in ascending order;

A3) for any pair (k,m) there exists a number $l \in \{1, \ldots, r\}$ such that $h_k H_1 h_m \neq 0$, $k, m = \overline{1,n}$, where h_k, $k = \overline{1,n}$ are linearly independent unit eigenvectors of H_0.

170   8 Control of Molecular and Quantum Systems

The following theorem, established in [19, 20] provides the mathematical basis for using the algorithm (8.3.15).

Theorem 8.1. Consider the system (8.3.12) with feedback control law (8.3.15), where Z and H_0 commute, the assumptions (A1), (A2), (A3) hold and $z_k < Z_* < z_{k+1}$. Then for any initial condition Ψ_0 from the set $M = \{\phi : z_k < \phi^* Z \phi < z_{k+1}\}$ the goal (8.3.13) is achieved.

The condition (A1) means that different values of the observable Z correspond to different pure states. The condition (A2) means that for different pairs of pure states the frequencies of transitions between them are different as well. Finally, the condition (A3) means that for any two pure states there exists a nonzero probability of transition between them for any (nonzero) control. For control of the molecule observables, the conditions imposed by Theorem 8.1 are not unnecessarily restrictive. For example, they are satisfied for the model of the hydrogen fluoride molecule discussed previously. The theorem remains true if a more general control algorithm is used $u = -F(\nabla_u \dot{Q}(t))$, where the function $F(x)$ is continuous, $F(x) = 0 \iff x = 0$, $F(x)x > 0$.

Note that since the control function is continuous and the sphere in $\mathbb{C}^n$ is compact and the theorem holds with arbitrarily small positive gain γ in (8.3.15), the goal is achieved for arbitrarily small intensity of control.

Some related results were also obtained by the method of Lyapunov functions, see [186, 187, 300, 301]. However, in [186, 300, 301] only the problem of stabilization (preparation) of the pure state was considered. Besides, the conditions A1–A3 are more mild than the conditions of [186, 300, 301] and allow for degeneracy of the energy levels. Particularly, it means that Theorem 8.1 applies to control of spin systems.

Let us apply the algorithm (8.3.15) to the problem of controlled dissociation of the hydrogen fluoride molecular ensemble. Algorithm (8.3.15) was applied to energy control of molecule HF, described by equation (8.2.11), (8.2.4).

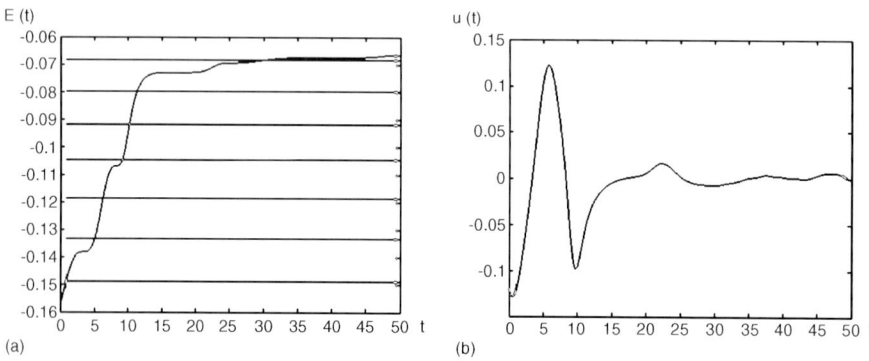

Fig. 8.3.3. (a) Energy evolution (horizontal lines are energy levels); (b) control function $u(t)$.

The constants for HF molecule are as follows [242]: $M = 1732$, $A = 0.4541$, $D = 0.2101$, $\alpha = 2.1350$, $r_0 = 1.75$ (Hartree units). The observable of interest is the energy H_0 with the average value $E(t) = \phi(t)^* H_0 \phi(t)$. Molecular energy levels are energies of pure states. The initial value of the phase vector is uniformly distributed between 3rd and 4th energy levels. The initial value of the energy is -0.1571 in atomic units. The goal value for energy is -0.06, it lies between 10th and 11th energy levels. The control function has the linear form $F(x) = Kx$ with the gain factor K equal to 200. The simulation time is 50 femtoseconds. The figures (Fig. 8.3.3) display evolution of energy ($E(t)$) and control function ($u(t)$). The simulation results confirm efficiency of the speed-gradient algorithm (8.3.15) for quantum control of diatomic molecule observables. More extensive simulations confirm that the set of initial conditions ensuring achievement of the control goal are broader then the one for which the theorem was proved.

9

Control Algorithms and Dynamics of Physical Systems

In this chapter, the links between control laws in technical systems and laws of dynamics in physical systems are examined. It is shown that the methods of control system design can be applied to the interpretation and explanation of dynamics laws for physical systems. The proposed approach is illustrated by the examples: motion of a particle in the potential field; wave, diffusion, and heat transfer equations; viscous flow equation. Applications to nonstationary nonlinear thermodynamics are presented: the derivation of the speed-gradient transient dynamics of the entropy maximization and an alternative proof of the extended Onsager principle.

9.1 Integral and differential variational principles

Consider a class of physical systems described by systems of differential equations

$$\dot{x} = f(x, u, t), \tag{9.1.1}$$

where $x \in \mathbb{R}^n$ is the vector of the system state, u is the vector of free (input) variables, $t \geq 0$. The problem of system evolution modeling can be posed as the search of a law of changing $u(t)$ in order to satisfy some criterion of "natural," or "reasonable" behavior of the system.

The equations of motion for either physical (natural) or technical (artificial) systems are often derived from variational principles (e.g., principle of least action, principle of least dissipation, maximum entropy principle) [168, 193, 252]. They are based on specification of a functional (usually, integral functional) and determination of real system motions $\{x(t), u(t)\}$ as points in an appropriate functional space providing the extrema of the specified functional. In order to explicitly describe either a control law or system dynamics the powerful calculus of variations or optimal control machinery is used.

In addition to integral principles, differential (local) ones were proposed: Gauss principle of least constraint, principle of minimum energy dissipation and others. It has been pointed out by Max Planck [353] that the local principles have some preference with respect to integral ones because they do not fix dependence of the current states and motions of the system on its later states and motions. One more local evolution principle is motivated by the speed-gradient method, see Section 2.4.2. It can be formulated as follows [135, 136, 146].

Speed-gradient (SG) principle. Among all possible motions only those are realized for which the input variables change proportionally to the speed gradient of an appropriate goal functional. If there are constraints imposed on the system motion, then the speed-gradient vector should be projected onto the set of admissible (compatible with constraints) directions.

In the next section the use of the speed-gradient principle will be illustrated by examples.

9.2 Examples of speed-gradient laws of dynamics

Suppose that the model (9.1.1) has a simple form

$$\dot{x} = u. \tag{9.2.2}$$

The relation (9.2.2) means that we are seeking for law of change of the state velocities. According to the speed-gradient principle, first we need to introduce the goal function $Q(x)$. The choice of $Q(x)$ should reflect the tendency of natural behavior to decrease the current value $Q(x(t))$.

Example 9.1. (Motion of a particle in the potential field). In this case the vector $x = \text{col}(x_1, x_2, x_3)$ consists of coordinates x_1, x_2, x_3 of a particle. Choose smooth $Q(x)$ as the potential energy of a particle and derive the speed-gradient law in the differential form. To this end, calculate the speed gradient

$$\dot{Q} = [\nabla_x Q(x)]^T u, \qquad \nabla_u \dot{Q} = \nabla_x Q(x).$$

Then, choosing the diagonal positive definite gain matrix $\Gamma = m^{-1} I_3$, where $m > 0$ is a parameter, I_3 is the 3×3 identity matrix, we arrive at familiar Newton's law $\dot{u} = -m^{-1} \nabla_x Q(x)$ or

$$m\ddot{x} = -\nabla_x Q(x). \tag{9.2.3}$$

Note that the speed-gradient laws with nondiagonal gain matrices Γ can be incorporated if a non-Euclidean metric in the space of inputs is introduced by the matrix Γ^{-1}. Admitting dependence of the metric matrix Γ on x one can obtain evolution laws for complex mechanical systems described by Lagrangian or Hamiltonian formalism.

9.2 Examples of speed-gradient laws of dynamics

The SG-principle applies not only to finite-dimensional systems, but also to infinite-dimensional (distributed) ones. Particularly, x may be a vector of a Hilbert space $\mathcal{X}$ and $f(x, u, t)$ may be a nonlinear operator defined in a dense set $D_F \subset \mathcal{X}$ (in such a case the solutions of (9.1.1) should be understood as generalized ones).

Example 9.2. (Wave, diffusion, and heat transfer equations). Let $x = x(r)$, $r = \mathrm{col}\,(r_1, r_2, r_3) \in \Omega$ be the temperature field or the concentration of a substance field defined in the domain $\Omega \subset \mathbb{R}^3$. Choose the goal functional as the following nonuniformity measure of the field

$$Q_t(x) = \frac{1}{2} \int_\Omega |\nabla_r x(r,t)|^2 \, dr, \qquad (9.2.4)$$

where $\nabla_r x(r,t)$ is the spatial gradient of the field. Assuming zero boundary conditions for simplicity, we have

$$\dot{Q}_t = -\int_\Omega \Delta x(r,t) u(r,t) \, dr, \qquad \nabla_u \dot{Q}_t = -\Delta x(r,t),$$

where $\Delta = \sum_{i=1}^{3} \frac{\partial^2}{\partial r_i^2}$ is the Laplace operator. Therefore, the speed-gradient evolution law in differential form is

$$\frac{\partial^2}{\partial t^2} x(r,t) = -\gamma \Delta x(r,t), \qquad (9.2.5)$$

which corresponds to the D'Alembert wave equation, while its finite form is

$$\frac{\partial x}{\partial t}(t) = -\gamma \Delta x(r,t) \qquad (9.2.6)$$

and coincides with the diffusion or heat transfer equation.

Note that the differential form of the speed-gradient laws corresponds to reversible processes while the finite form generates irreversible ones.

Example 9.3. (Viscous flow equation). Let $v(r,t) \in \mathbb{R}^3$ be the velocity field of fluid, $p(r,t)$ be the pressure field, i.e., $x = \mathrm{col}\,(v(r,t), p(r,t))$. Introduce the goal functional as follows

$$Q_t = \int_\Omega p(r,t) \, dr + \nu \int_\Omega |\nabla_r v(r,t)|^2 \, dr, \qquad (9.2.7)$$

where $\nu > 0$. Calculation of the speed-gradient with respect to (9.2.2) yields $\nabla_u \dot{Q}_t = \nabla_r p - \nu \Delta v$. Then, the differential form of speed gradient is just the Navier–Stokes equation for viscous fluid motion

$$\rho \frac{\partial v}{\partial t}(r,t) = -\nabla_r p(r,t) + \nu \Delta v(r,t), \qquad (9.2.8)$$

where $\nu > 0$ is the viscosity coefficient, $\rho = \gamma^{-1}$ is density.

176   9 Control Algorithms and Dynamics of Physical Systems

Other examples of reproducing dynamical equations for mechanical, electrical, and thermodynamic systems can be found in [135]. The SG-principle applies to a broad class of physical systems subjected to potential and/or dissipative forces. On the other hand, nonpotential systems with presence of vortex motions (e.g., mechanical systems affected by gyroscopic forces) cannot be derived by the SG-method.

9.3 Speed-gradient entropy maximization

It is worth noticing that the speed-gradient principle provides an answer to the question: *how* the system will evolve? It differs from the principles of maximum entropy, maximum Fisher information, etc. providing and answer to the questions: *where?* and *how far?* Particularly, it means that SG-principle generates equations for the transient (nonstationary) mode rather than the equations for the steady-state mode of the system. It allows one to study nonequilibrium and nonstationary situations, stability of the transient modes, maximum deviations from the limit mode, etc. Let us illustrate this feature by example of entropy maximization problem.

It was mentioned in Section 1.4 that, according to the 2nd thermodynamics law and to the Maximum Entropy Principle of Gibbs–Jaynes the entropy of any physical system tends to increase until it achieves its maximum value under constraints imposed by other physical laws. Such a statement provides knowledge about the final distribution of the system states, i.e., about asymptotic behavior of the system when $t \to \infty$. However, it does not provide information about the way how the system moves to achieve its limit (steady) state.

In order to provide motion equations for the transient mode we employ the SG-principle. Assume for simplicity that the system consists of N identical particles distributed over m cells. Let N_i be the number of particles in the ith cell and the mass conservation law hold:

$$\sum_{i=1}^{m} N_i = N. \tag{9.3.9}$$

Assume that the particles can move from one cell to another and we are interested in the system behavior both in the steady-state and in the transient modes. The answer for the steady-state case is given by the Maximum Entropy Principle: if nothing else is known about the system, then its limit behavior will maximize its entropy. Define, according to the deterministic style of this book, the entropy of the system as logarithm of the number of possible states:

$$S = \ln \frac{N!}{N_1! \cdots N_m!}. \tag{9.3.10}$$

If there are no other constraints except normalization condition (9.3.9) it achieves maximum with $N_i^* = N/m$. For large N an approximate expression

9.3 Speed-gradient entropy maximization

is of use. Namely, if the number of particles N is large enough, one may use the Stirling approximation $N_i! \approx (N_i/e)^{N_i}$. Then

$$S \approx N \ln \frac{N}{e} - \sum_{i=1}^{m} N_i \ln \frac{N_i}{e} = N \ln N - \sum_{i=1}^{m} N_i \ln N_i = -\sum_{i=1}^{m} N_i \ln \frac{N_i}{N}$$

which is proportional to the standard stochastic entropy $S = -\sum_{i=1}^{m} p_i \ln p_i$, if the probabilities p_i are understood as frequencies N_i/N.

To get the answer for the transient mode let us apply the SG-method choosing the approximate entropy $\hat{S}(X) = N \ln N - \sum_{i=1}^{m} N_i \ln N_i$ as the goal function to be maximized, where $X = \text{col}(N_1, \ldots, N_m)$ is the state vector of the system (here the *state* is understood in the systems theory sense). Assume for simplicity that the motion is continuous in time and the numbers N_i are changing continuously, i.e., N_i are not necessarily integer (for large N_i it is not a strong restriction). Then the sought law of motion can be represented in the form

$$\dot{N}_i = u_i, \ i = 1, \ldots, m, \tag{9.3.11}$$

where $u_i = u_i(t)$, $i = 1, \ldots, m$ are controls – auxiliary functions to be determined. According to the SG-method one needs to evaluate first the speed of change of the entropy (9.3.10) with respect to the system (9.3.11), then the speed-gradient (gradient of the speed with respect to the vector of controls u_i considered as frozen parameters) and finally define actual controls proportionally to the projection of the speed-gradient to the surface of constraints (9.3.9). In our case the goal function is the entropy S and its speed coincides with the entropy production $\dot{S}$. In order to evaluate $\dot{S}$ let us again approximate S from the Stirling formula $N_i! \approx (N_i/e)^N$:

$$\hat{S} = N \ln N - N - \sum_{i=1}^{m}(N_i \ln N_i - N_i) = N \ln N - \sum_{i=1}^{m} N_i \ln N_i. \tag{9.3.12}$$

Evaluation of $\dot{\hat{S}}$ yields

$$\dot{\hat{S}} = -\sum_{i=1}^{m}\left((u_i \ln N_i + N_i \frac{u_i}{N_i}\right) = -\sum_{i=1}^{m} u_i(\ln N_i + 1).$$

It follows from (9.3.9) that $\sum_{i=1}^{m} u_i = 0$. Hence $\dot{\hat{S}} = -\sum_{i=1}^{m} u_i \ln N_i$. Evaluation of the speed-gradient yields $\frac{\partial \dot{\hat{S}}}{\partial u_i} = -\ln N_i$ and the SG-law is as follows:

$$u_i = \gamma(-\ln N_i + \lambda), \ i = 1, \ldots, m, \tag{9.3.13}$$

where Lagrange multiplier λ is chosen in order to fulfill the constraint $\sum_{i=1}^{m} u_i = 0$, i.e., $\lambda = \frac{1}{m}\sum_{i=1}^{m} \ln N_i$. The final form of the system dynamics law is as follows:

$$\dot{N}_i = \frac{\gamma}{m} \sum_{i=1}^{m} \ln N_i - \gamma \ln N_i, \quad i = 1, \ldots, m. \tag{9.3.14}$$

According to the SG-principle the equation (9.3.14) determines transient dynamics of the system. To confirm consistency of the choice (9.3.14) let us find the steady-state mode, i.e., evaluate asymptotic behavior of the variables N_i. To this end note that in the steady-state $\dot{N}_i = 0$ and $\sum_{i=1}^{m} \ln N_i = \ln N_i$. Hence, all N_i are equal: $N_i = N/m$ which corresponds to the maximum entropy state and agrees with thermodynamics.

The next step is to examine stability of the steady-state mode. It can be done by means of the entropy Lyapunov function

$$V(X) = S_{\max} - S(X) \geq 0, \tag{9.3.15}$$

where $S_{\max} = N \ln m$. Evaluation of $\dot{V}$ yields

$$\dot{V} = -\dot{S} = \sum_{i=1}^{m} u_i \ln N_i = \frac{\gamma}{m} \Big[\Big(\sum_{i=1}^{m} \ln N_i \Big)^2 - m \sum_{i=1}^{m} (\ln N_i)^2 \Big].$$

It follows from the Cauchy–Bunyakovsky–Schwarz inequality that $\dot{V}(X) \leq 0$ and the equality $\dot{V}(X) = 0$ holds if and only if all the values N_i are equal, i.e., only at the maximum entropy state. Thus the law (9.3.15) provides global asymptotic stability of the maximum entropy state. The physical meaning of the law (9.3.15) is moving along the direction of the maximum entropy production rate (direction of the fastest entropy growth).

Let in addition to the mass conservation law (9.3.9) the energy conservation law hold. What can be said about transient and steady-state modes? The case of more than one constraint can be treated in the same fashion. Let E_i be the energy of the particle in the ith cell and the total energy $E = \sum_{i=1}^{m} N_i E_i$ be conserved. The energy conservation law

$$E = \sum_{i=1}^{m} N_i E_i \tag{9.3.16}$$

appears as an additional constraint. Acting in a similar way, we arrive at the law (9.3.14) which needs modification to ensure conservation of the energy (9.3.16). According to the SG-principle one should form the projection onto the surface (in our case – subspace of dimension $m-2$) defined by the relations

$$\sum_{i=1}^{m} u_i E_i = 0, \quad \sum_{i=1}^{m} u_i = 0. \tag{9.3.17}$$

It means that the evolution law should have the form

$$u_i = \gamma(-\ln N_i) + \lambda_1 E_i + \lambda_2, \quad i = 1, \ldots, m, \tag{9.3.18}$$

where λ_1, λ_2 are determined by substitution of (9.3.18) into (9.3.17). The obtained equations are linear in λ_1, λ_2 and their solution is given by formulas

$$\begin{cases} \lambda_1 = \frac{\gamma m \sum_{i=1}^{m} E_i \ln N_i - \gamma (\sum_{i=1}^{m} E_i)(\sum_{i=1}^{m} \ln N_i)}{m \sum_{i=1}^{m} E_i^2 - (\sum_{i=1}^{m} E_i)^2}, \\ \lambda_2 = \frac{\gamma}{m} \sum_{i=1}^{m} \ln N_i - \frac{\lambda_1}{m} \sum_{i=1}^{m} E_i. \end{cases} \quad (9.3.19)$$

The solution of (9.3.19) is well defined if $m \sum_{i=1}^{m} E_i^2 - (\sum_{i=1}^{m} E_i)^2 \neq 0$ which holds unless all the E_i are equal (degenerate case).

Let us evaluate the equilibrium point of the system (9.3.11), (9.3.18) and analyze its stability. At the equilibrium point of the system the following equalities hold: $\gamma(-\ln N_i) + \lambda_1 E_i + \lambda_2 = 0$, $i = 1, \ldots, m$. Hence

$$N_i = C \exp(-\mu E_i), \quad i = 1, \ldots, m, \quad (9.3.20)$$

where $\mu = \lambda_1/\gamma$ and $C = \exp(-\lambda_2/\gamma)$.

The value of C can also be chosen from the normalization condition $C = N(\sum_{i=1}^{m} \exp(-\mu E_i))$. We see that equilibrium of the system with conserved energy corresponds to the Gibbs distribution which agrees with classical thermodynamics. Again it is worth to note that the direction of change of the numbers N_i coincides with the direction of the fastest growth of the local entropy production subject to constraints.

As before, it can be shown that (9.3.15) is Lyapunov function for the system and that the Gibbs distribution is the only stable equilibrium of the system in nongenerate cases. Therefore again the SG-principle allows to easily determine transient modes of the system. It complements the framework of classical thermodynamics.

Similar results are valid for continuous (distributed) systems even for more general problem of minimization of relative entropy (Kullback divergence). In this case to evaluate the speed of change of the goal function one needs to use the Pavon–Ticozzi formula [341].

9.4 Onsager relations

The speed-gradient approach provides a new insight for various physical facts and phenomena. For example, we will give evidence for an extended version of the symmetry principle for kinetic coefficients (Onsager principle) in thermodynamics [178, 318, 323] (it is also called the Maxwell–Betti theorem in elasticity theory). Consider an isolated physical system whose state is characterized by a set of variables (thermodynamic parameters) $\xi_1, \xi_2, \ldots, \xi_n$. Let $x_i = \xi_i - \xi_i^*$ be deviations of the variables from their equilibrium values $\xi_1^*, \xi_2^*, \ldots, \xi_n^*$. Let the dynamics of the vector $x_1, x_2, \ldots, x_n$ be described by the differential equations

$$\dot{x}_i = u_i(x_1, x_2, \ldots, x_n), \quad i = 1, 2, \ldots, n. \quad (9.4.21)$$

Linearize equations (9.4.21) near equilibrium

$$\dot{x}_i = -\sum_{k=1}^{n} \lambda_{ik} x_k, \quad i = 1, 2, \ldots, n. \tag{9.4.22}$$

The *Onsager principle* [178] claims that the values λ_{ik} (so-called kinetic coefficients) satisfy the equations

$$\lambda_{ik} = \lambda_{ki}, \quad i, k = 1, 2, \ldots, n. \tag{9.4.23}$$

In general, the Onsager principle is not valid for all systems or far from equilibrium. Its existing proofs (see, e.g., [256]) require additional postulates. Below the new proof is given, showing that it is valid for irreversible speed-gradient systems without exceptions.

First of all, the classical formulation of the Onsager principle (9.4.23) should be extended to nonlinear systems. A natural extension is the following set of identities:

$$\frac{\partial u_i}{\partial x_k}(x_1, x_2, \ldots, x_n) = \frac{\partial u_k}{\partial x_i}(x_1, x_2, \ldots, x_n). \tag{9.4.24}$$

Obviously, for the case when the system equations (9.4.21) have linear form (9.4.22) the identities (9.4.24) coincide with (9.4.23). However, since linearization is not used in the formulation (9.4.24) there is a hope that the extended version of the Onsager law holds for some nonlinear systems far from equilibrium. The following theorem specifies a class of systems for which this hope comes true.

Theorem 9.1. *Assume that there exists a smooth function $Q(x)$ such that equations (9.4.21) represent the speed-gradient law in finite form for the goal function $Q(x)$.*

Then, the identities (9.4.24) hold for all $x_1, x_2, \ldots, x_n$.

Proof of Theorem 9.1. The proof is very simple. Since (9.4.21) is the speed-gradient law for $Q(x)$, its right-hand sides can be represented in the form

$$u_i = -\gamma \frac{\partial \dot{Q}}{\partial u_i}, \quad i = 1, 2, \ldots, n.$$

Therefore, $u_i = -\gamma(\partial Q/\partial x_i)$ (in view of $\dot{Q} = (\nabla_x Q)^T u$). Hence

$$\frac{\partial u_i}{\partial x_k} = -\gamma \frac{\partial^2 Q}{\partial x_i \partial x_k} = \frac{\partial u_k}{\partial x_i},$$

and identities (9.4.24) are valid. ∎

Thus, for speed-gradient systems the extended form of the Onsager equations (9.4.24) hold without linearization, i.e., they are valid not only near the equilibrium state. It is worth mentioning that the above derivation is valid only under the assumption that all the derivatives exist, i.e., all the involved functions are smooth. It excludes a number of nonsmooth physical problems, like description of shock waves. In a special case the condition (9.4.24) was proposed in [129].

Note that the condition (9.4.24) is necessary and sufficient for potentiality of the vector-field of the right-hand sides of (9.4.21), i.e., existence of a scalar function $\bar{Q}$ such that $u_i = \gamma \nabla_x \bar{Q} = \gamma \nabla_u \dot{Q}$. It means that generalized Onsager relations (9.4.24) are necessary and sufficient for the thermodynamics system to obey the SG-principle for some $\bar{Q}$. On the other hand, it is known that different potential functions for the same potential vector-field can differ only by a constant: $\bar{Q} = Q + \text{const}$ and their stationary sets coincide.

It means, in turn, that if the system tends to maximize its entropy and at every time instant it tends to maximize its entropy production rate (Prigogine principle) then the generalized Onsager principle (9.4.24) holds and vice versa. Indeed, in the case the entropy can serve as the goal function for the speed-gradient evolution law. Note that for special case the relation between Prigogine principle and Onsager principle was established by D. Gyarmati [192].

For the speed-gradient systems some other properties can be established. Let, for example, a system is governed by SG-law with a convex entropy goal function S. Then the decrease of the entropy production $\dot{S}$ readily follows from the identities

$$\ddot{S} = d\dot{S}/dt = (\nabla_x \dot{S})^T \dot{x} = \gamma (\nabla_x ||\nabla_x S||^2)^T \nabla_x S = 2\gamma (\nabla_x S)^T [\nabla_x^2 S](\nabla_x S).$$

If the entropy $S(x)$ is convex then its Gessian matrix $\nabla_x^2 S$ is negative semidefinite: $\nabla_x^2 S \leq 0$. Hence $\ddot{S}(x) \leq 0$ and $\dot{S}$ cannot increase [135].

9.5 Discussion: Dynamics and the purpose

It is shown in this chapter that the SG-principle previously used for control system design, is applicable to the interpretation and explanation of dynamics laws for physical systems. The SG-principle belongs to a family of extremal (variational) principles and its peculiarity is in active using the concept of the *goal*. Although using extremal principles is by no means a new approach, most of previous applications belong to the engineering area where optimality is a goal of creating an artificial engineering system. In the contrast, the goal-seeking in physics was many times criticized as a way of scientific description of nature.

It is interesting to link the speed-gradient approach to the views of Howard Rosenbrock [377–379] who demonstrated how to obtain the Schrödinger's equation and some other results from the elementary theory of quantum mechanics by means of the optimality principle of dynamic programming. In [379], Rosenbrock quotes Albert Einstein [125]:

> For the scientist there is only "being," but no wishing, no valuing, no good, no evil; no goal.

Rosenbrock characterizes such an opinion as outdated, arguing that the goal-seeking is natural for much broader class of systems than just living organisms. His arguments are as follows [379]:

> As an example, living organisms exhibit clear purposes, and if the substrate of quantum mechanical particles from which life evolves is described as purposeless the question arises "how can purpose arise from a purposeless substrate?"

The speed-gradient principle described in this chapter as a local (differential) extremal principle relies upon the goal-seeking idea even more heavily than integral variational principles and it may add arguments into the discussion. In the cases where obtaining physical results is easier from extremal principle than from system equations (see [379]), using simple speed-gradient formulation may further facilitate analysis of a physical system and provide further insights about its properties. It may be especially helpful in the cases when the idea of the goal is intrinsic to the system and formation of a goal function does not look artificial. By the way, it correlates with one of the fundamental biological principles: organisms and populations evolve in such way that the rate of increase of their biomass is maximized [421]. Therefore, it may be helpful in development of unifying views of both animated and inanimate nature.

10
Examples

10.1 Controlled Stephenson–Kapitsa pendulum

In the beginning of the 20th century, Scottish professor P. Stephenson has shown mathematically that the upper unstable equilibrium of a mathematical pendulum can be stabilized by fast vibration of its suspension point [417]. In about 40 years, in the end of the 1940s Russian physicist, future Nobel prize winner Piotr Kapitsa surprised his colleagues by experiment with a rod eccentrically mounted on a horizontal motor shaft. The demonstration showed that the upper unstable equilibrium of the swinging rod (pendulum) can be made stable by sufficiently fast vibrations of the pivot. The experimental results were explained both by Kapitsa himself who developed his method of "effective potential" [224] (see also [71, 73]) and by mathematician Nikolai Bogoliubov by means of the method of averaging (history and explanations see, e.g., in [72]). The above mentioned and other results started the development of a new field in mechanics called "Vibrational mechanics" with numerous applications in science and technology [73]. Similar ideas formed the basement of a corresponding branch of the control theory: *vibrational control* [62, 292]. It is important to stress that Kapitsa's experiment was, perhaps, the first one clearly demonstrating the possibility and physical consequences of changing properties of a physical system by means of control.

Let us revisit Kapitsa's experiment from the feedback point of view. The mathematical model of Kapitsa's pendulum differs from the one mentioned in Chapter 2 in that the controlling action is the vertical acceleration of the pivot rather than the applied torque, see Fig. 10.1.1. Therefore, the model of the system under control is as follows

$$J\ddot{\varphi} + \varrho\dot{\varphi} + mgl\sin\varphi = mlu\sin\varphi, \qquad (10.1.1)$$

where $\varphi = \varphi(t)$ is the angle of deflection of the pendulum from its lower vertical position; $u = u(t)$ is the vertical acceleration of the suspension point; $J = ml^2$ is the moment of inertia of the pendulum; $\varrho \geq 0$ is the friction

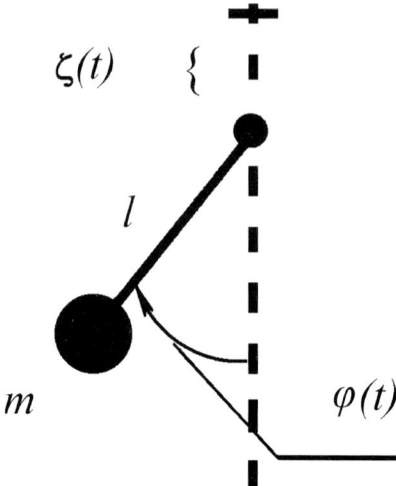

Fig. 10.1.1. Pendulum with vibrating pivot.

coefficient. Since acceleration of the pivot is proportional to the applied force, we assume that it plays a role of the controlling variable. Kapitsa studied behavior of the pendulum under harmonic law of the pivot motion with the frequency ω and the amplitude A. For that case control function $u(t)$ has the form

$$u(t) = A\omega^2 \sin \omega t. \qquad (10.1.2)$$

Kapitsa's experiments discovered an effect of stabilization of the pendulum near the upper unstable equilibrium. Numerous theoretical investigations both before and after Kapitsa's work show that stabilization of the unstable equilibrium occurs at sufficiently large excitation frequency ω, i.e., in the case when control function (10.1.1) is large [73, 224]. In this case the changes of the pendulum pivot position may remain small, strengthening paradoxicalness of the phenomenon. However, the magnitude of the applied controlling force required for stabilization of the upper equilibrium by a high-frequency excitation should be large.

Now, let us address the following question: Is it possible to achieve a similar behavior of the pendulum (10.1.1) by applying control $u(t)$ of smaller magnitude if it is allowed to use feedback laws for vibration of the pivot?

A conventional approach of automatic control theory to find a control law is based on linearization of the controlled system model. However, in our case linearization does not work. Indeed, linearization provides good approximation of the model only near an equilibrium position or near a reference trajectory, i.e., initial conditions should be close to such a trajectory. In our case no reference trajectory of the pendulum is specified and the problem is to examine a global behavior of the pendulum for all initial conditions.

10.1 Controlled Stephenson–Kapitsa pendulum

Let us introduce an auxiliary control goal

$$\lim_{t\to\infty} H(t) = H_*, \tag{10.1.3}$$

where

$$H = \frac{J}{2}(\dot\varphi)^2 + mgl(1 - \cos\varphi) \tag{10.1.4}$$

is the total energy of the pendulum. The goal (10.1.3) differs from conventional control goals (regulation and tracking). It rather reminds the goal of a man swinging the swings or the goal of a monkey swinging the liana. Similar goals may also be posed when design of a walking robot, a pendulum clocks, etc.

A common sense suggests that one needs much less power to swing the pendulum than to fix it in a given position, if not in an equilibrium. So, the question is: Is it possible to swing the swings up to the upper position by means of small control?

Let us apply the speed-gradient method, choosing the goal function as the square error between the current and the desired energy values: $Q = (H - H_*)^2/2$, where $H_* = 2mgl$ is the energy of the upper equilibrium. Evaluate the speed of changing the value of Q along trajectories of the system (10.1.1) with frozen u, and then evaluate the gradient of the speed with respect to control. In our case control is scalar and the gradient is just single partial derivative in u. We arrive at simple algorithms

$$u = -\gamma(H - H_*)\dot\varphi\sin\varphi, \tag{10.1.5}$$

$$u = -\gamma\,\text{sign}\left[(H - H_*)\dot\varphi\sin\varphi\right]. \tag{10.1.6}$$

Let us choose the algorithm (10.1.6). Then it follows from Theorem 4.2 (see Example 4.1) that the energy level achievable in the system (10.1.1), (10.1.6) is not less than

$$\overline{H} = \frac{1}{2}\left(\frac{\gamma}{\varrho}\right)^2. \tag{10.1.7}$$

Therefore, the energy level $H_* = 2mgl$ will be achieved for $\gamma > 2\varrho\omega_0$. Particularly, for $\varrho = 0$, the stabilization of the energy surface $H = H_*$ is achieved for arbitrarily small amplitude of control γ. Moreover, if the damping ϱ is small, than the control magnitude γ may be chosen small too.

Achievement of the desired energy level does not imply stabilization of any equilibrium point belonging to this level. However, it was shown in the papers [407, 408] that in the case $\varrho = 0$, the algorithm (10.1.6) for $H_* = 2mgl$ ensures convergence $H(\varphi(t),\dot\varphi(t)) \to H_*$ and convergence $(\varphi(t),\dot\varphi(t)) \to (\pi, 0)$ when $t \to \infty$ at any initial conditions except lower equilibrium. Besides, the control magnitude $\gamma > 0$ can be chosen arbitrarily small. Although in the case $\rho > 0$ no stabilization of all (or almost all) trajectories near the points $(\pm\pi, 0)$ in conventional sense can be observed, the portion of trajectories staying near $(\pm\pi, 0)$ tends to unity as $t \to \infty$ (see Fig.10.1.2). Therefore, stabilization of the upper equilibrium occurs in statistical sense.

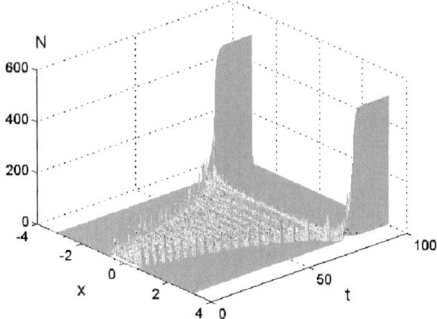

Fig. 10.1.2. Distribution of the pendulum deflection angle for the pendulum (10.1.1) affected by control (10.1.6) for $J = 1, \varrho = 0.025, m = 1, l = 1, g = 9.81, \gamma = 0.1$. Gistogramme is made for 1000 trajectories, initial distribution is uniform on $[-\pi/4,\ \pi/4]$.

The problem of stabilizing the pendulum by motion of the suspension point has the following interesting feature. Since we consider acceleration of the suspension point $u(t)$ as control, it follows from general properties of speed-gradient algorithms that $u(t) \to 0$ when $t \to \infty$. However, nothing definite can be concluded about the behavior of the speed and position of the suspension point. Formal model may admit that the speed of the suspension point and the deflection of its position from initial value do not approach zero, and even may increase infinitely. Of course, such a behavior does not have either physical or practical meaning.

Let us now describe a modification of the control algorithm, free from above-mentioned drawback [165]. To this end, introduce an extended goal function

$$Q_1 = Q + \frac{1}{2}z^\mathrm{T} P z, \qquad (10.1.8)$$

where $z = \mathrm{col}(\zeta, \dot\zeta)$, $P = P^\mathrm{T} \geq 0$ is a positive semidefinite weighting matrix, ζ, $\dot\zeta$ are height and speed of the suspension point, respectively. Then the relation $\ddot\zeta = u$ can be interpreted as an additional equation of motion, i.e., the system turns into a system with two degrees of freedom and the state $x = \mathrm{col}(\varphi, \dot\varphi, \zeta, \dot\zeta)$.

According to the speed-gradient method, evaluation of the speed of changing the goal function yields

$$\dot Q_1 = \dot Q + z^\mathrm{T} P \begin{bmatrix}\dot\zeta\\0\end{bmatrix} + z^\mathrm{T} P \begin{bmatrix}0\\1\end{bmatrix} u, \qquad (10.1.9)$$

where $\nabla_u \dot Q_1 = (H_0 - H_*)\dot\varphi \sin\varphi + p_{22}\dot\zeta + p_{12}\zeta$, p_{11}, p_{22} are the elements of the second column of the matrix P. Therefore, the modified algorithm looks as follows:

$$u = -\gamma(H_0 - H_*)\dot\varphi \sin\varphi - \mu\dot\zeta - \nu\zeta, \qquad (10.1.10)$$

where $\gamma > 0, \mu > 0, \nu > 0$ are control gains.

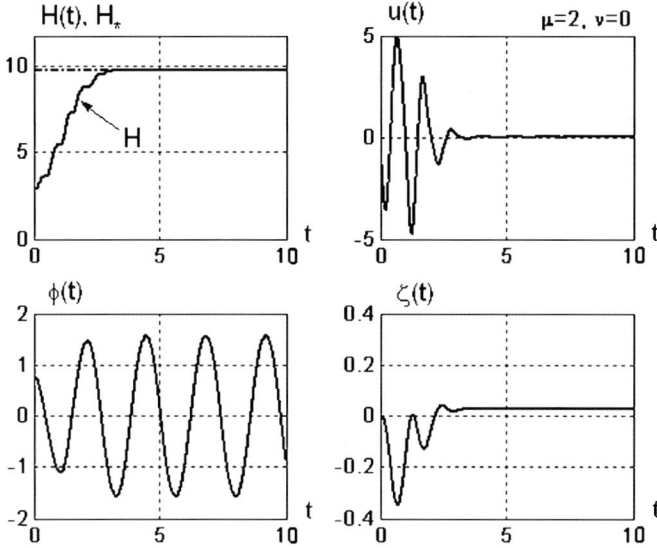

Fig. 10.1.3. Simulation of the pendulum with the algorithm (10.1.10) for $\mu = 2$ and $\nu = 0$.

The results of Chapter 3 are not applicable to the obtained system since the initial controlled system is not Hamiltonian. Nevertheless, employing more general results about the properties of the speed-gradient control [157, 165], one can show that the new control goal is achieved and $\zeta(t) \to$ const for almost all initial conditions, if $\nu = 0$. Simulation results of the system with control law (10.1.10) for $m = 1, l = 1, H_* = mgl = 9.81, \gamma = 0.7, \mu = 2, \nu = 0$ are shown in Fig. 10.1.3.

Additionally, if we choose $\mu > 0$, $\nu > 0$, then the closed-loop system possesses a more strong property $\zeta(t) \to 0$, i.e., deflection of the suspension point from its initial position asymptotically vanishes. Such a behavior is illustrated by Fig 10.1.4, where $\nu = 2$, and other parameters are same as in the previous case.

In a similar way one may solve the swinging control problems for the cases when the suspension point is moving along horizontal or inclined line, rather than vertically. However, more complex cases may require more efforts to cope with. For example, additional difficulties may be caused by incomplete or noisy measurements, e.g., in the case when the angular velocity $\dot{\varphi}(t)$ is not available for measurement. In such cases one may need to introduce additional filters (observers) into the system. Another difficulty may arise due to incompleteness of control, e.g., one cannot neglect inertia properties of the motor driving the pendulum. In this case the dynamics of the controlled system are described by equations

$$J\ddot{\varphi} + mgl\sin\varphi = mlu\sin\varphi, \quad T\dot{u} + u = v, \qquad (10.1.11)$$

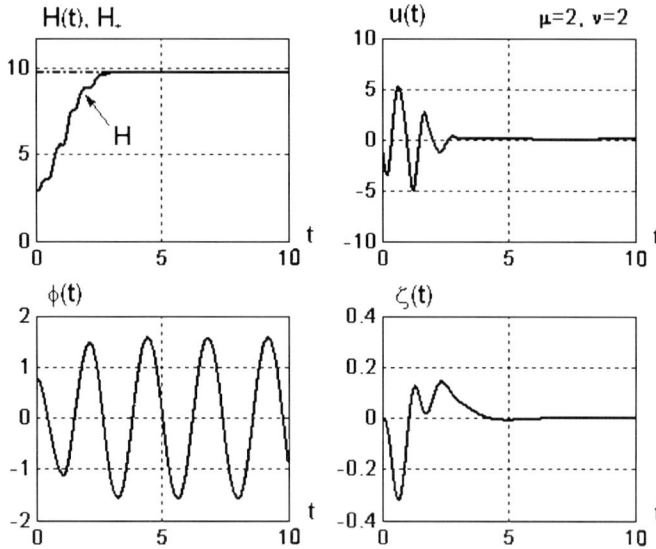

Fig. 10.1.4. Simulation of the pendulum with the algorithm (10.1.10) for $\mu = 2$ and $\nu = 2$.

where $v = v(t)$ is a new control signal. The problem in this case is that the control variable $v(t)$ does not appear in the right-hand side of the first equation (10.1.11) and direct calculation of the speed-gradient yields zero.

Fortunately, the modern theory of nonlinear and adaptive control provides a broad arsenal of methods to overcome the above-mentioned difficulties as well as many other ones, see [157, 164, 245].

Other examples of advantageous properties of feedback are provided in the literature on control of chaos where highly unstable orbits are shown to be stabilizable by tiny corrections, see Chapter 6 and [27, 103, 164, 403]. It is also worth to notice that the swinging of the pendulum has become a kind of benchmark example both in mechanics and in control literature [8, 45, 69, 107, 138]. An advantage of the speed-gradient method used in this section is its applicability to more general, higher-dimensional systems, see Chapter 7.

10.2 Escape from a potential well and lossless communications

The study of escape from a potential well is important in many fields of physics and mechanics [253, 385, 436]. Sometimes escape is an undesirable event and it is important to find conditions preventing it (e.g., buckling of the shells, capsize of the ships, etc.). In other cases escape is useful and the conditions guaranteeing it are needed. Escape may correspond to a phase transition in the system. For crystalline lattices an escape corresponds to a dislocation,

10.2 Escape from a potential well and lossless communications

see Chapter 7, while in the area of information physics briefly described in Chapter 1, escape may correspond to transition from the state "0" to state "1" of the information system, i.e., to creation of a bit of information. In all cases the conditions of achieving escape by means of as small external force as possible are of interest. Usually, such conditions are obtained by extensive computer simulations for the case when external force is a periodic and even a harmonic function of time. Below, following [140, 147] we study properties of escape achieved by means of a feedback excitation.

Consider nonlinear oscillators with one degree of freedom, modeled as

$$\ddot{\varphi} + \varrho\dot{\varphi} + \Pi(\varphi)' = u, \qquad (10.2.12)$$

where $\varrho > 0$ is the damping coefficient. Equation (10.2.12) can be transformed to the Hamiltonian form with coordinate and momentum $q = \varphi$, $p = \dot{\varphi}$, the Hamiltonian function (energy) $H_0(\varphi, \dot{\varphi}) = \frac{1}{2}\dot{\varphi}^2 + \Pi(\varphi)$ and passivity output $\dot{\varphi}$.

In [418], such a problem (optimal escape) has been studied for typical nonlinear oscillators (10.2.12) with a single-well potential $\Pi_e(\varphi) = \varphi^2/2 - \varphi^3/3$ (so-called "escape equation") and a twin-well potential $\Pi_D(\varphi) = -\varphi^2/2 + \varphi^4/4$ (Duffing oscillator). The least amplitude of a harmonic external forcing $u(t) = \bar{u}\sin\omega t$ for which no stable steady-state motion exists within the well was determined by intensive computer simulations. For example, for escape equation with $\varrho = 0.1$ the optimal amplitude was evaluated as $\bar{u} \approx 0.09$, while for Duffing twin-well equation with $\varrho = 0.25$ the value of amplitude was about $\bar{u} \approx 0.21$. We performed computer simulations for the case of Duffing oscillator. The results agree with those of [418]. The typical time histories of input and output for $\bar{u} = 0.208$ are shown in Fig. 10.2.5(a). It is seen that escape does not occur.

Using feedback forcing we may expect reducing the escape amplitude. In fact, using the formula (4.2.43), the amplitude of feedback (4.2.40) or (4.2.41) leading to escape can be easily calculated, just substituting the height of potential barrier $\max_\Omega \Pi(\varphi) - \min_\Omega \Pi(\varphi)$ for $\overline{H}$ into (4.2.43) where Ω is the well corresponding to the initial state. For example, in the case of escape equation $\overline{H} = 1/6, \varrho = 0.1$, and $\bar{u} = 0.0577$, while for Duffing oscillator with $\overline{H} = 1/4, \varrho = 0.25$ escape amplitude is estimated as $\bar{u} = 0.1767$. The obtained values are substantially smaller than those evaluated in [418] for harmonic excitation. The less the damping, the bigger the difference between the amplitudes of feedback and nonfeedback signals leading to escape. Simulation exhibits still stronger difference: escape for Duffing oscillator occurs for $\bar{u} = \gamma = 0.122$ if the feedback (4.2.41) or (4.2.44) is applied, see Fig. 10.2.5(b). Note that the oscillations in the feedback systems have both variable frequency and variable shape.

We also studied the dependence of escape amplitudes on the damping by means of computer simulations in the range of damping coefficient ϱ varying from 0.01 to 0.25. Simulations confirmed theoretical conclusion that the feedback escape amplitude is proportional to the damping, see Fig. 10.2.6(a). We

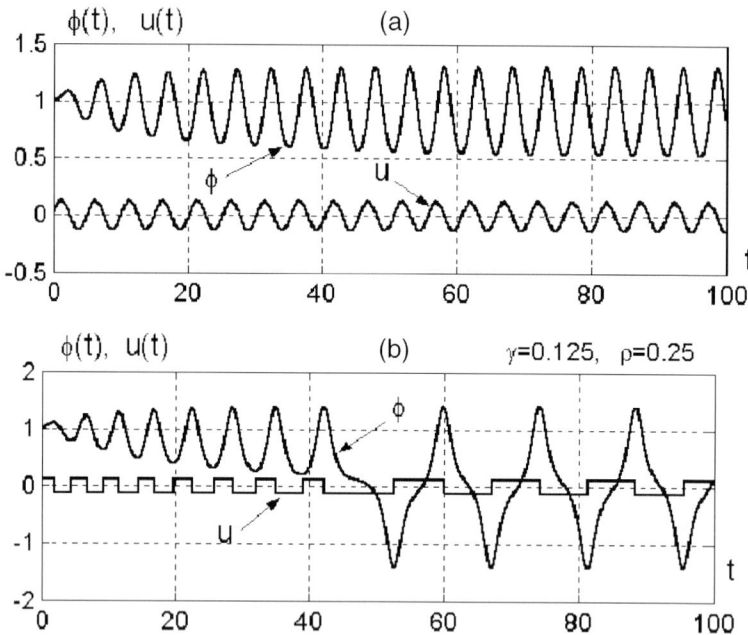

Fig. 10.2.5. Escape from a potential well for the Duffing system. (a) Harmonic excitation; (b) speed-gradient excitation.

may evaluate the efficiency of feedback μ as the ratio of escape amplitudes for harmonic ($\bar{u}_h$) and feedback ($\bar{u}_f$) forcing:

$$\mu = \frac{\bar{u}_h}{\bar{u}_f}. \qquad (10.2.13)$$

Figure 10.2.6(b) shows that the efficiency of feedback is inversely proportional to the damping for small values of damping.

The feedback approach can possibly be applied to implementation of "almost reversible" communication proposed by Landauer [258, 260]. It was pointed out in [258, 260] that in order to transmit a bit of information one may just change the state of a bistable device and transport it. The physical origin of a device does not matter: it may be a mechanical relay, magnetic domain, or a molecule. Information storage and transmission with molecular size devices is of special interest for design of molecular computers. A number of attempts to implement elementary acts of information processing with atoms and molecules have been reported recently [333, 420, 426]. As suggested in [420], the ammonia (NH_3) molecule can be a promising candidate for achieving molecular control due to a number of its useful properties, e.g., the isolated ammonia molecule can be confined in a fullerene without any large interactions with the inner wall. It also has two stable states, see Fig. 10.2.7.

10.2 Escape from a potential well and lossless communications 191

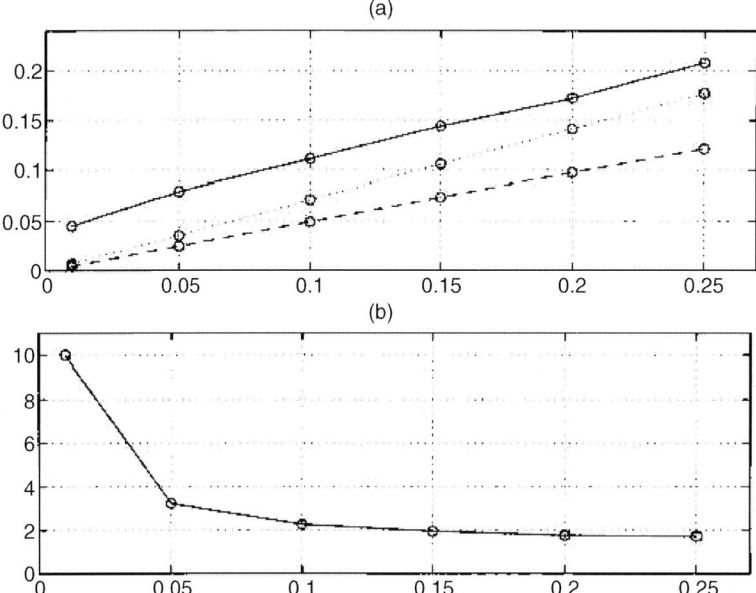

Fig. 10.2.6. Efficiency of feedback versus excitation degree for escape from a Duffing potential well. (a) Level of control action required for escape (A – harmonic excitation; B – feedback excitation by means of speed-gradient algorithm, estimate evaluated numerically; C – feedback excitation by means of speed-gradient algorithm, estimate evaluated analytically from Theorem 4.1). (b) Efficiency of feedback: ratio of (A) over (C).

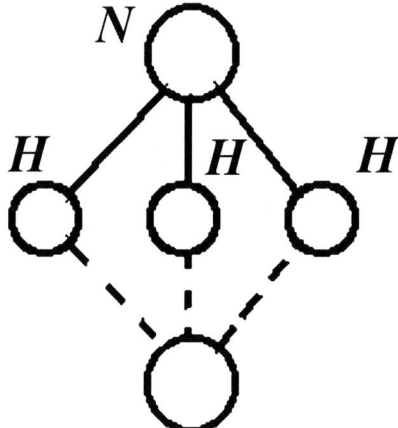

Fig. 10.2.7. Two configurations of ammonia molecule.

The previously described method of controlled escape from a potential well can be used for energy saving information processing at a molecular level. To design laser pulses performing transition of a molecule from one potential well to another one an approach developed in Chapter 8 can be employed. The two wells may correspond, for example, to upper and lower positions of N atom in the NH_3 molecule. The approach of Chapter 8 suggests to design a proper laser pulse by means of computer simulation of the closed loop speed-gradient control system for a single molecule (so-called reference molecule). Then the designed pulse (controlling function $u(t)$) is applied to a real physical system in a nonfeedback manner. Efficiency of such an approach has been demonstrated for a number of molecular tasks [18, 19, 21, 149, 155]. Perspectives of usage these results in real molecular computer applications depend on progress in laser technologies (development of blue and UV lasers) aimed at achieving the lateral size of the laser beam compatible with the size of a single molecule or a small group of molecules. The proposed approach could provide and efficient and realizable alternative to the technique of "deapening" and "lawering" of a bistable potential proposed in [258, 260].

10.3 Feedback spectroscopy

The conventional spectroscopy is based upon applying a harmonic signal to the physical system under examination. Though the energy eigenvalues in the spectroscopy theory are predicted by quantum mechanical calculations, to explain the dynamics of resonant interaction between radiation and matter the classical harmonic oscillator model is usually used [119]. Real multi-DOF system has a variety of natural modes with different natural frequencies and different losses. The most interesting are the resonant frequencies, corresponding to small damping which produce the resonant peaks (lines) on the spectrogram. The resonant peaks can be evaluated by scanning over the frequency range of input signal. What is the role of nonlinearity?

The conventional methods treat anharmonicity as perturbation changing resonance conditions for large deviations from equilibrium. As a result some energy is reflected instead of being absorbed by the system and the energy value (10.1.1) cannot be achieved for larger $\bar{u}$.

Let us try feedback. Applying the signal of form (4.2.40) and using the nonlinear oscillator model (10.2.12) we can achieve the energy level (4.2.43) coinciding with (10.1.1). Thus we get an opportunity of giving full degree of excitation to the system and evaluating its energy absorbing ability at higher energy levels. Since the nonlinearity is essential only for small damping ϱ, i.e., near linear resonances, the "feedback" spectroscopy techniques should incorporate the conventional ones in order to determine the near resonant regions and to give initial excitation to the system.

It is important that for excitation we may use simple feedback (4.2.41) which does not require measuring energy and looks like just introduction of a negative damping into the system. Therefore, the obtained resonant energy value does not depend on the shape of potential, i.e., the kind of anharmonicity does not matter.

Of course the feedback excitation is not easy to implement because it should depend not only on the intensity but also on the phase of the radiation. However, the development of ultrafast controlled lasers [460], growth of the computers productivity, and increase of measurements speed and accuracy give hope for the experimental verification of the approach. It is already quite realistic for the fields dealing with lower frequencies, e.g., for ultrasonic investigations. Another approach to nonlinear resonance spectroscopy which does not use energy considerations was suggested in [100].

10.4 Control of chemical reaction with phase transition

The discovery of oscillatory Belousov–Zhabotinsky chemical reaction in the 1950s has drawn an interest in oscillatory behavior in chemistry. Usage of oscillatory and even chaotic regimes opens new horizons in chemical technology. Recently, the possibility of an oscillatory mode in a chemical reaction with phase transition was established by S. Kukushkin and A. Osipov [246, 247]. The models suggested in [246, 247] can be applied to the deposition of thin films with a chemical reaction. Regulation of the desired oscillatory regime (amplitude or frequency) is important to produce a film with the desired structure and properties. However, to design a good controller one needs to overcome the difficulties caused by nonavailability of some variables for measurement and uncertainty in the system model.

In this section, following [151, 164] we employ the method of adaptive control based on linearization of the Poincaré map and solving the goal inequalities discussed in Section 6.4.2 to control a chemical reaction with phase transition.

10.4.1 Problem statement

Consider a chemical reaction of the type $A + B \leftrightarrow C$. Let the rate of the reaction be equal to kAC^2, where A and C denote the concentrations of the corresponding substances, and k is the reaction constant (the inverse reaction rate is assumed to be zero). As it was shown by Kukushkin and Osipov [246] the chemical reaction and nucleation processes are described by the simplified model

$$\begin{cases} dA/dt = J_0 - kAC^2 \\ dC/dt = kAC^2 - \gamma NC \\ dN/dt = \beta_0(C - C_e), \end{cases} \quad (10.4.14)$$

where J_0 is the arrival rate of component A at a substrate (evaporation of A and C from the substrate is neglected), N is the concentration of new phase islands, C_e is the equilibrium concentration, γ and β_0 are constants of proportionality. We introduce dimensionless constants $J = kJ_0/\beta_0\gamma$ and $y_0 = C_e k^{2/3}\beta_0^{-1/3}\gamma^{-1/3}$ and new dimensionless variables as follows

$$x = Ak^{2/3}\beta_0^{-1/3}\gamma^{-1/3},$$
$$y = Ck^{2/3}\beta_0^{-1/3}\gamma^{-1/3},$$
$$z = Nk^{1/3}\beta_0^{-2/3}\gamma^{1/3},$$
$$\tau = tk^{-1/3}\beta_0^{2/3}\gamma^{2/3}$$

Then the system (10.4.14) reads

$$\begin{cases} dx/d\tau = J - xy^2 \\ dy/d\tau = xy^2 - yz \\ dz/d\tau = y - y_0 \end{cases} \quad (z \geq 0). \tag{10.4.15}$$

Computer simulations show that in the case when the constant flow J is less than $J_1 \approx 0.888$ (with a precision of 0.0002), then the system behaves in an ordinary way (Fig. 10.4.8); that is, $C(t)$ tends to C_e, $N(t)$ tends to $J_0/\gamma C_e$, $A(t)$ tends to $J_0/\gamma C_e^2$ when $t \to \infty$ and the growth of the film continues due to an increase of the nucleus sizes. With $J_1 < J < J_2 \approx 1.049$ such behavior becomes locally unstable, which results in oscillations and functions $C(t), N(t), A(t)$ turn out to be periodic (a stable limit cycle, Fig. 10.4.9). For $J > J_2$ this limit cycle disappears and the film grows in the unstable "accumulative" mode (Fig. 10.4.10).

In this situation an important problem is control of the size of oscillations because the structure and properties of the film depend on the amplitude and the period of oscillations. Particularly, the following question is of interest: how to change the external flow $J(t)$ in order to make the local maximum concentration of new phase islands in the oscillating mode $z_{\max}$ close to the given value z_*.

The control goal can be formalized as follows

$$|z_k - z_*| \leq \Delta, \quad \Delta > 0, \tag{10.4.16}$$

where $z_k = z(t_k)$, t_k is the time when $z(t)$ achieves its kth local maximim, Δ is the given accuracy threshold. Since $n = 3$, the linearized discrete model has the second-order form

$$z_{k+1} + a_1 z_k + a_2 z_{k-1} = b_0 J_k + b_1 J_{k-1} + \varphi_k \tag{10.4.17}$$

where J_k is the control action at the kth step, a_1, a_2, b_0, b_1 are unknown coefficients, φ_k is the bounded disturbance.

10.4.2 Adaptive control algorithm

The adaptive control algorithm includes the main loop algorithm which calculates the new value of the control action J_k, and the adaptation algorithm

10.4 Control of chemical reaction with phase transition 195

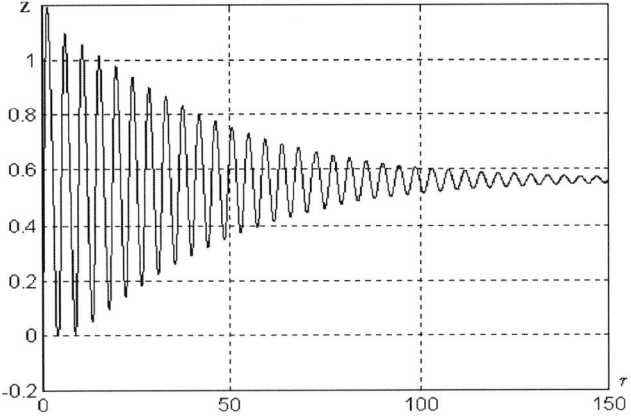

Fig. 10.4.8. Plot of $z(t)$ versus t for the normal growth mode ($J = 0.7$).

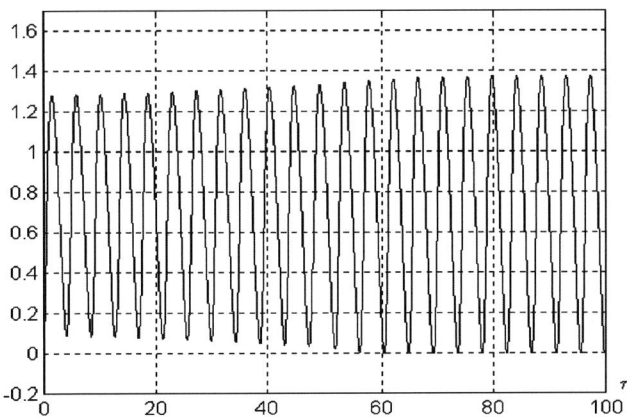

Fig. 10.4.9. Plot of $z(t)$ versus t for the oscillatory growth mode ($J = 0.9$).

which updates the estimates of the system (10.4.17) parameters $\hat{a}_{1k}, \hat{a}_{2k}$, $\hat{b}_{0k}, \hat{b}_{1k}$. According to the results of Section 6.4.2, the main loop algorithm can be designed as follows

$$J_k = [z_* + \hat{a}_{1k} z_k + \hat{a}_{2k} z_{k-1} - \hat{b}_{1k} J_{k-1}]/\hat{b}_{0k}. \quad (10.4.18)$$

This algorithm is chosen to provide achievement of the goal in one step if the estimates coincide with the true plant model parameters. The parameter update algorithm adjusting the estimates is chosen by the method of the goal inequalities and takes the following form

$$\begin{aligned}
\hat{a}_{i,k+1} &= \hat{a}_{i,k} - \alpha \vartheta_k z_{k-i+1}, & i &= 1, 2, \\
\hat{b}_{i,k+1} &= \hat{b}_{i,k} - \alpha \vartheta_k J_{k-i}, & i &= 0, 1, \\
\vartheta_k &= \begin{cases} z_{k+1} - z_*, & |z_{k+1} - z_*| > \Delta, \\ 0, & |z_{k+1} - z_*| \leq \Delta, \end{cases}
\end{aligned} \quad (10.4.19)$$

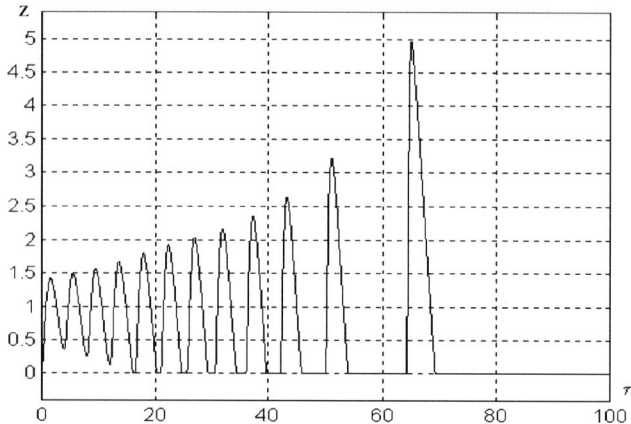

Fig. 10.4.10. Plot of $z(t)$ versus t for the unstable accumulative growth mode ($J = 1.2$).

where $\alpha > 0$ is the adaptation gain.

It follows from the results of Section 6.4.2 that if $|\varphi_k| \le \Delta_\varphi < \Delta$ and the value of α is small enough, then the control goal (10.4.16) is achieved after a finite number of steps, i.e., the relation (10.4.16) holds for any $k > k_*$ for some k_*.

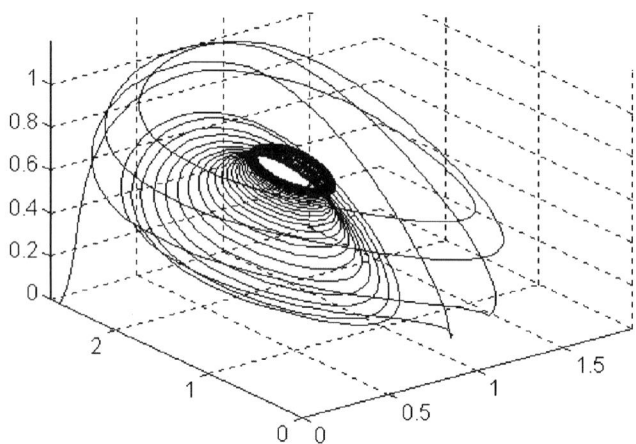

Fig. 10.4.11. Phase plot for $z_* = 0.9$.

10.4.3 Simulation results

Investigation of accuracy and convergence rate of the algorithm (10.4.19) was performed by computer simulation. In Figs. 10.4.11, 10.4.12, and 10.4.13 show

10.4 Control of chemical reaction with phase transition   197

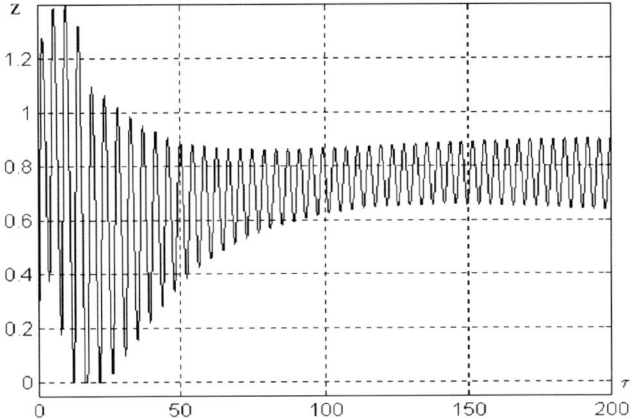

Fig. 10.4.12. Plot of $z(t)$ versus t for $z_* = 0.9$.

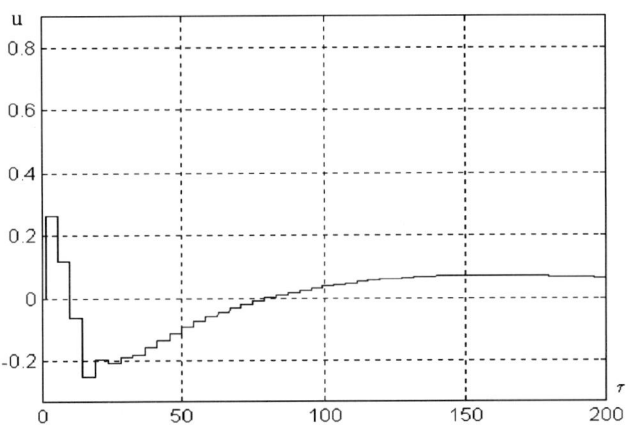

Fig. 10.4.13. Plot of $J(t)$ versus t for $z_* = 0.9$.

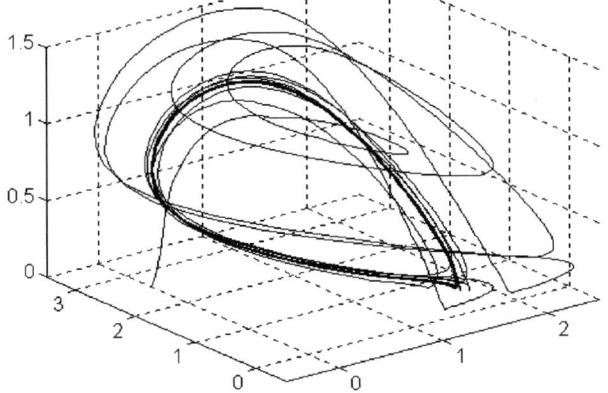

Fig. 10.4.14. Phase plot for $z_* = 1.5$.

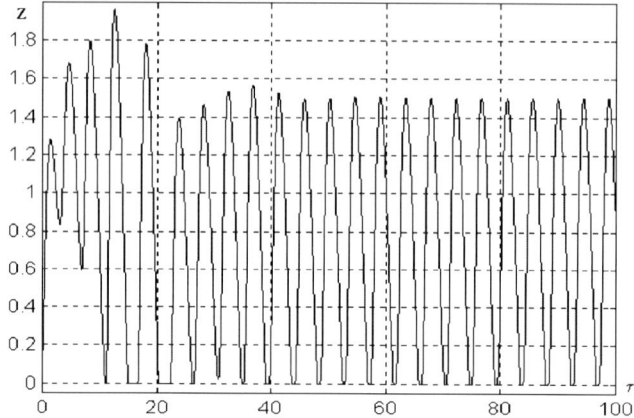

Fig. 10.4.15. Plot of $z(t)$ versus t for $z_* = 1.5$.

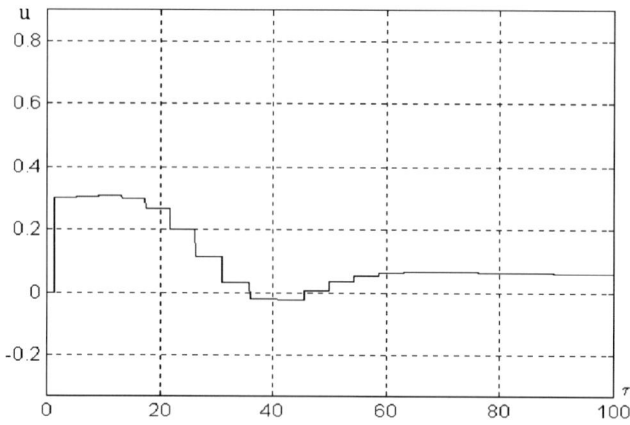

Fig. 10.4.16. Plot of $J(t)$ versus t for $z_* = 1.5$.

the phase plot and the plots of $z(t)$ and $J(t)$ versus t for the control goal (10.4.16) with $z_* = 0.9$. The following initial conditions and parameter values were chosen: $J(0)=0.9$; $x(0)=0$; $y(0)=2.6$; $z(0)=0$; $y_0=5/4$. It is seen that $J(t)$ is close to 0.96 at large time values. Note that the amplitude of new phase island concentration oscillations decreases significantly and becomes approximately twice less than the amplitude of oscillations for the constant $J(t)=0.96$. Figures 10.4.14, 10.4.15, and 10.4.16 show the similar results for the case $z_*=1.5$. As would be expected, in this case the amplitude of oscillations increases and also $J(t) \to 0.96$ when $t \to \infty$. This confirms the efficiency of the proposed method.

The numerous simulations discovered some phenomena that need further investigation. For example, it has been seen that the system behavior at large time values depends not only on the asymptotic values of the external parameters, but also on the way of their changing at the earlier stages of the process. This effect is analogous to the frequency capture phenomenon in the oscillating system with two degrees of freedom, which is well known in the oscillations theory [253]. It has been also noticed that using more complex discretized model of the third-order instead of (10.4.17) allows one to increase the convergence rate of adaptation. Additionally, a small change in the initial values of the adaptation parameters may lead to a significant change of the limit cycle without violation of the control goal.

10.5 Energy-like control of predator–prey system

In this section, we study a problem of controlling oscillations in a population consisting of two interacting species, following [164]. We employ the simplest and most popular model of population dynamics, so-called Lotka–Volterra model. Consider a system of two species (preys and predators) interacting with each other. Let the populations be large, so that the number of individuals can be treated as real numbers.

Let $x_1(t)$ be the number of preys measured at time t and $x_2(t)$ be the number of predators measured at the same time. In the absence of predators ($x_2 = 0$) the population of preys increases infinitely ($\dot{x}_1 = ax_1$), where $a > 0$ is the birth rate of preys. On the other hand, predators need preys to survive. Therefore, in the absence of preys the predators starve to death ($\dot{x}_2 = -dx_2$), where $d > 0$ is the death rate of the predators.

The famous Lotka–Volterra model is described by the following differential equations:
$$\begin{cases} \dot{x}_1 = ax_1 - bx_1x_2 \\ \dot{x}_2 = -dx_2 + cx_1x_2 \end{cases} \quad (10.5.20)$$

where $b > 0, c > 0$. Thus, if the number of predators is large the number of preys decreases while if the number of preys is small the number of predators decreases as well, and the system (10.5.20) has infinite number of periodic motions (cycles) [380].

Introduce the controlled version of the model (10.5.20), assuming that the birth rate of preys can be controlled. Then the model (10.5.20) is modified as follows:
$$\begin{cases} \dot{x}_1 = ax_1 - bx_1x_2 + x_1u \\ \dot{x}_2 = -dx_2 + cx_1x_2 \end{cases} \quad (10.5.21)$$

where u is the control action.

The uncontrolled system ($u(t) \equiv 0$) has the following *first integral*:

$$H(x_1, x_2) = \left(cx_1 - d - d\log\frac{cx_1}{d}\right) + \left(bx_2 - a - a\log\frac{bx_1}{a}\right) \quad (10.5.22)$$

Indeed, easy calculations show that

$$\dot H(x_1, x_2) = 0$$

along any solutions of (10.5.20) ($x_1(0) > 0, x_2(0) > 0$) that means that the quantity H preserves constant value. The first integral (10.5.22) can be interpreted as a "total energy" of the "predator–prey" system and the control goal can be posed in terms of achieving of a desired level of the quantity H:

$$H(x_1(t), x_2(t)) \to H^* \quad \text{as} \quad t \to \infty \qquad (10.5.23)$$

where $H^* \geq 0$ is the desired level of the first integral. The less the value H^*, the smaller the size of the corresponding periodic cycle.

Similarly to many of previous examples, we use the speed-gradient method to design the control algorithm. To this end, introduce the following objective function Q:

$$Q(x_1, x_2) = \frac{1}{2} (H(x_1, x_2) - H^*)^2$$

Evaluation of its time derivative with respect to system (10.5.21) yields

$$\dot Q(x_1, x_2, u) = (H(x_1, x_2) - H^*)(cx_1 u - du).$$

Evaluating gradient in u we get

$$\frac{\partial Q}{\partial u}(x_1, x_2) = (H(x_1, x_2) - H^*)(cx_1 - d)$$

and arrive at the following speed-gradient control algorithm

$$u(t) = -\gamma (H(x_1(t), x_2(t)) - H^*)(cx_1(t) - d), \quad \gamma > 0 \qquad (10.5.24)$$

Applying the results of the Section 3.5, it is easy to show that the goal (10.5.23) is achieved for arbitrary $\gamma > 0$ and for all initial conditions $x_1(0) > 0, x_2(0) > 0$. The proof is based on the Lyapunov function

$$V(x_1, x_2) = Q(x_1, x_2)$$

which decreases along any solution of the controlled system. It is worth mentioning that the choice $H^* = 0$ corresponds to the control algorithm which asymptotically stabilizes the system (10.5.21) in its inherent equilibrium ($x_{10} = d/c, x_{20} = a/b$), and the limit cycle collapses to a point.

Theoretical results are illustrated by computer simulation for the following predator–prey model:

$$\begin{cases} \dot x_1 = 4x_1 - 2x_1 x_2 + x_1 u \\ \dot x_2 = x_1 x_2 - x_2 \end{cases} \qquad (10.5.25)$$

This system has nonisolated periodic solutions which correspond to the following first integral

10.5 Energy-like control of predator–prey system

$$H(x_1, x_2) = (x_1 - 1 - \log x_1) + \left(2x_2 - 4 - 4\log \frac{x_2}{2}\right).$$

The phase trajectory of the system (10.5.25) corresponding to the initial conditions $x_1(0) = 4, x_2(0) = 3$ is shown in Fig. 10.5.17.

The control algorithm (10.5.24) for the system (10.5.25) has the following form

$$u(t) = -\gamma \left(H(x_1(t), x_2(t)) - H^*\right)(x_1(t) - 1)$$

The results of computer simulation are presented in Fig. 10.5.18 for the following parameters $\gamma = 1, x_1(0) = 3, x_2(0) = 3$. Plots (a), (b), (c) show the transient behavior which corresponds to the desired level of H equal to 0,1,3, respectively. In the picture Fig. 10.5.18 (d) transient processes in "energy" are shown. It is seen that choosing various values of the desired "energy" level H^* we can achieve various oscillatory behavior of the controlled system.

The proposed approach can be extended to the systems of many interacting species with populations $x_1, \ldots, x_n$, described by controlled version of the equations [380]

$$\begin{cases} \dot{x}_i = x_i(d_i - \sum_{j=1}^{n} a_{ij}x_j + b_i(x)u_i, i = 1, \ldots, n. \end{cases} \qquad (10.5.26)$$

The speed-gradient method is designed for the goal function $Q(x) = (H(x) - H^*)^2$, where $H(x)$ has the form

$$H(x_1, \ldots, x_n) = \sum_{i=1}^{n} x_i^* \left(\frac{x_i}{x_i^*} - d\ln \frac{x_i}{x_i^*}\right), \qquad (10.5.27)$$

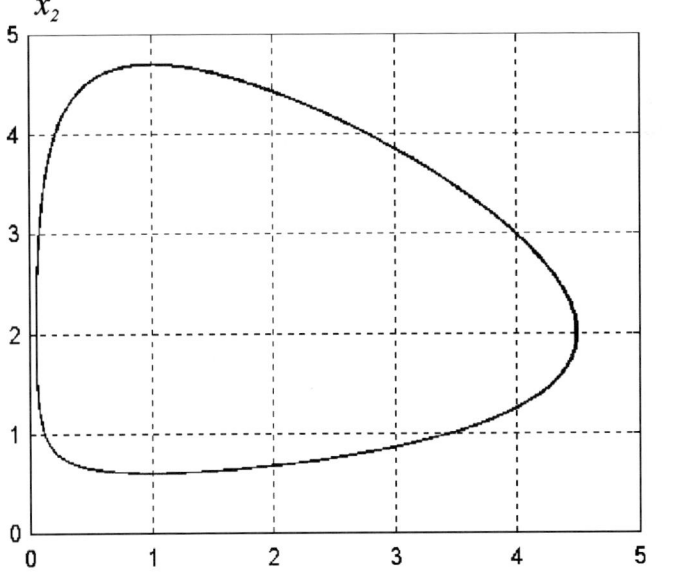

Fig. 10.5.17. Phase portrait of the uncontrolled predator–prey system

where $x^* = A^{-1}D$ is the vector of the system nontrivial equilibrium. It is known [380] that $H(x)$ is the first integral of the system if $x_i^* > 0$. If $b(x) > 0$ than the speed-gradient algorithm

$$u(t) = -\gamma\left(H(x) - H^*\right)b(x)$$

is well defined and ensures the achievement of the goal $H(x(t)) \to H^*$.

10.6 Control of noise-induced transition

It is well known that the behavior of many nonlinear dynamical systems is influenced by noise. Noise may cause a number of interesting phenomena which cannot be observed in noiseless systems. A characteristic example is the nonlinear stochastic resonance [32, 199, 254]. In some cases the noise plays the constructive role, e.g., sensitivity of detection of weak signals by some systems can be notably increased due to the presence of noise [199]. Besides, nonlinear oscillators may undergo a phase transition at the presence of noise. That is, there can be observed a qualitative change of the dynamic of a system when the amplitude of noise reaches some critical value [458]. The well-known example is the emergence of oscillations of a large amplitude due to noise in a system that is stable at the absence of noise. The oscillations appear as the

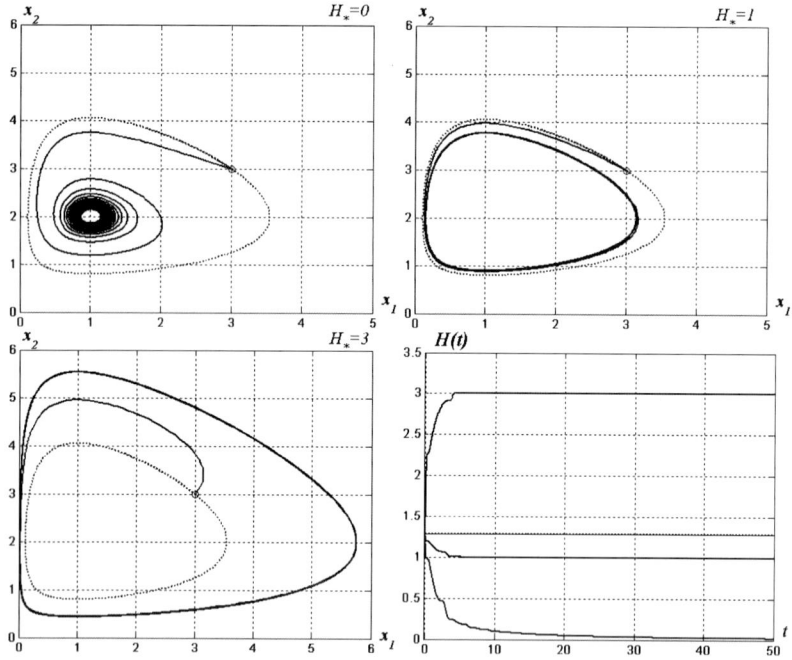

Fig. 10.5.18. Transient behavior of the controlled predator–prey system.

response of the system to the modulation of some of the system's parameter by noise. In this case, the macroscopic dynamic of the system to a large extent is determined by noise (which, in a real system, originates from microscopic processes). Such a phenomenon – parametric noise resonance – is, essentially, a variant of the nonlinear stochastic resonance.

The appearance of the parametric noise resonance is undesirable when it destroys the normal functioning of a dynamic system. As an example, this may take place in converters of optical images that are used for recording of high speed processes in the optical infrared range [39]. The operation of the converters is based on electronic properties of the structure "semiconductor – gas-discharge gap." It has been found that, at a small current density, the intrinsic noise of the device can initiate the oscillations of such a large amplitude that the gas discharge in the gap is spontaneously interrupted [39]. In other words, there is observed the transition from a conductive state of the system to the dielectric one. In a dielectric state, the device becomes insensitive to an incoming pulse of light and is not able, therefore, to convert an image.

It is of the practical importance to investigate, whether it is possible to provide the proper control of the considered (and similar) systems while applying a low amplitude action that varies in time. The purpose of such a control is to suppress the escape of the dynamic system from the area of the phase space where the main function of the system is ensured.

Below a realization of such an approach is demonstrated, following [44]. The nonlinear model of the device introduced in [39] (Astrov model) is used for control design and simulation. Control is provided by the proper variations of one of the parameters of the model. In the real device, it corresponds to the feeding voltage. To design the feedback control algorithm, the speed-gradient method is again implemented. The dynamic of the system at the presence of control is analyzed in the work by numerical solving of the corresponding equations.

For the parameters chosen in the present analysis, the positive result of control is obtained at a rather small value of the control amplitude that can equal only a small fraction of the main feeding voltage. It is revealed that the efficiency of control depends, in a nonmonotonic way, on the amplitude of the control parameter, and exemplifies a resonance-like regularity. In our opinion, this indicates relation of the obtained results to the nonlinear stochastic resonance phenomenon, the essence of which is the nonmonotonic response of a system to the deterministic signal as dependent on the amplitude of stochastic (noise) input.

10.6.1 The system model and problem statement

We study the model of the device dynamics proposed by Yu. Astrov [39]:

$$dE/dt = a_1(E_m - E) - b_1 N E , \qquad (10.6.28)$$

$$dN/dt = \frac{N}{\tau}\left(\frac{E}{E_c} - 1\right), \qquad (10.6.29)$$

where E is the electric field strength in the discharge gap and N is the density of free charge carriers in the gap. The first equation describes the charging of the capacity of the discharge gap from a source of the feeding voltage and its discharging due to the presence of free carriers in the gap. The characteristic time of the charging process is $\tau_E = 1/a_1$, b_1 is a coefficient. The maximum value of E in the gap that can be provided by the source of constant voltage, is E_m. For its value we have the obvious relationship $E_m = U_m/d$, where U_m is the voltage of the feeding source and d is the length of the gap in the direction of the electric current flow.

The second equation describes the dynamics of density of free carriers in the gap which is governed by processes of their generation and decay. It is supposed that the process of carriers generation prevails over their recombination when E value is larger than some critical electric field E_c. The parameter τ defines the rate of temporal variation of charge carriers density when the electric field in the gap is not equal to the critical value.

The Astrov model (10.6.28, 10.6.29) has been successfully used for interpretation of some peculiarities in dynamics of the experimental devices [39–41], such as the appearance of oscillations at low current density and the spontaneous interruption of the discharge glow. The latter effect can be identified as the noise-induced transition. When such a transition is realized, the converter cannot process an incoming image.

In the absence of noise, the stationary solution of the system (E_0, N_0)

$$E_0 = E_c, \qquad N_0 = \frac{a_1}{b_1}\left(\frac{E_m}{E_c} - 1\right) \qquad (10.6.30)$$

is stable. The linear analysis of stability of the solution reveals that it is the stable focus for large enough values of τ_E. If τ_E increases further, the oscillatory properties of the system become more pronounced: that is, its Q-factor grows. In its turn, the value of N_0 decreases which means the lowering of electrical current in the device modeled with (10.6.28), (10.6.29).

It should be stressed that generation of free carriers in the gas discharge gap is provided by the avalanche ionization of gas atoms and molecules. The efficiency of this process is known to fluctuate in time, which serves as a source of the intrinsic noise of the experimental nonequilibrium system. In a simple approach, the influence of the intrinsic noise on dynamics of the system can be modeled with using (10.6.28), (10.6.29) when stochastically changing the parameter E_c in time [39]. It has been found in the cited work that such an approach can give the growth of oscillations in time and be the reason of interruption of the discharge glow in the device.

It is of interest to study whether it is possible, by introducing a proper control of the amplitude of the feeding voltage, to suppress the tendency of the system to find itself in the dangerous domain of the phase space. The next section is devoted to the elaboration of an algorithm for such a control.

10.6.2 Control algorithm design

First of all note that physical principles of the device in question suggest that the role of controlling action can be played either by supply voltage or conductance of the semiconductor component. The latter option can be implemented in an experiment due to the photoelectrical effect in the semiconductor – by means of its optical excitation. By applying such a method, one can vary the coefficient a_1 in the system model (10.6.28).

In this example, the first option is adopted and the device is controlled by changing the supply voltage. It is assumed that all the state variables $E(t)$, $N(t)$ of the model (10.6.28), (10.6.29) are available for measurement. To construct the control algorithm, the speed-gradient method is used.

The first step of the speed-gradient method procedure is the choice of the goal function $Q(x)$ where small values correspond to achievement of the control goal. Since the control goal is to maintain the system trajectories near the nominal mode (10.6.30), the goal function can be chosen as square norm of the deflection $\Delta X = X(t) - X_0$, where $X(t) = col[E(t), N(t)]$, $X_0 = col[E_0, N_0]$. To simplify design we use the linearized model of (10.6.28), (10.6.29). The linearization near X_0 yields

$$\dot{X} = A(X - X_0), \qquad (10.6.31)$$

where

$$A = \begin{bmatrix} -a_1 - b_1 N_0 & -b_1 E_0 \\ N_0/(\tau E_c) & 0 \end{bmatrix}.$$

Obviously, A is a stable (Hurwitz) matrix if $a_{11} > 0$ and $-a_{21} * a_{12} > 0$. In our case, these inequalities are valid.

Now let us choose the goal function as a quadratic form: $Q(X) = (X - X_0)^T P(X - X_0)$, where P is the symmetric positive definite 2x2 matrix that is the solution to the Lyapunov equation $PA + A^T P = -I$. Let the control variable be $u = E_m - E_{mnom}$. Then one needs to evaluate the partial derivative of dQ/dt in control which is $dQ/dt = -(X - X_0)^T(X - X_0)$. Taking into account that control $u(t)$ appears both in X and X_0, yields

$$\frac{\partial \dot{Q}}{\partial u} = \mu - \eta u,$$

where $\mu = a_1 p_{11}(E - E_0) + a_1 p_{21}(N - N_0) - \dfrac{a_1 p_{21}}{b_1 E_0} \cdot$

$\cdot [a_1(E_m - E) - b_1 N E]$

$$\eta = 2\frac{a_1^2}{b_1 E_0} p_{21}$$

Choosing the speed-gradient algorithm in the finite form, see Section 2.4.2, we obtain the following control algorithm:

$$u = -\frac{\gamma\mu}{1-\gamma\eta}, \qquad (10.6.32)$$

where $\gamma > 0$ is the design parameter (gain). It is convenient to analyze a modification of (10.6.32) having the fixed amplitude of control:

$$u = -\bar{u}\,\mathrm{sat}\left(\frac{\gamma\mu}{\bar{u}(1-\gamma\eta)}\right), \qquad (10.6.33)$$

where $\bar{u} > 0$ is the new design parameter,

$$\mathrm{sat}(z) = \begin{cases} z, & |z| \leq 1, \\ \mathrm{sign}\,z, & |z| \geq 1. \end{cases}$$

If the deviation ΔX is small, and the noise is absent, the convergence of the algorithm (the achievement of the control goal) follows from the stability of the linearized system. Otherwise, the behavior of the closed control loop system (10.6.28), (10.6.29), (10.6.33) is studied by means of computer simulation.

10.6.3 Control system analysis

The goal of the analysis is to study the possibility to suppress the noise-induced transition by means of the control mechanism introduced above and to examine the quantitative dependence of the efficiency of the control on the amplitude of the control action.

For simulation, the following values of the model (10.6.28),(10.6.29) parameters that correspond to parameters of the real device [39, 40], were chosen: $a_1 = 10^4$, $b_1 = 5 \times 10^{-3}$, $\tau = 1.5 \times 10^{-9}$, $E_m = 8 \times 10^4$, $E_c = E_0 = 4 \times 10^4$, $N_0 = 2 \times 10^6$. The amplitude of the uniformly distributed stochastic noise was 1% of E_c.

The first stage of the analysis is to study the dynamics of the uncontrolled system at the presence of noise while varying the main parameter of the system (a_1). The problem is to determine how the value a_1 influences the time t_c of the transition of the system into the nonconductive (dielectric) state. In the present calculations, this time is specified as the time of the first crossing of the level $N_* = 100$ by the trajectory $N(t)$ (Remark that the value $N_* = 100$ is in the correspondence to the minimum discharge current in the device: Reaching of the state with the carriers density less than $N_* = 100$ would mean to find less than one electron in the gap, which is, in reality, a nonconducting case. So, the discharge interrupts when the state $N_* = 100$ is reached). The found dependence of the time before discharge interruption t_c on a_1 is shown in Fig. 10.6.19. Based on the obtained results, the value $a_1 = 10^4$ was chosen as the starting value for further experiments.

The next step is to study a possibility to maintain the discharge by means of control. Since the system trajectories are random functions, the minimum value of the discharge current $N_* = min\ N(t)$ is chosen as the measure of the

10.6 Control of noise-induced transition   207

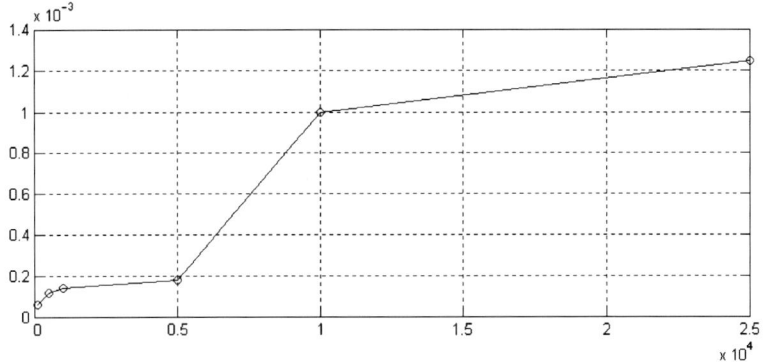

Fig. 10.6.19. Plot of t_c versus a_1.

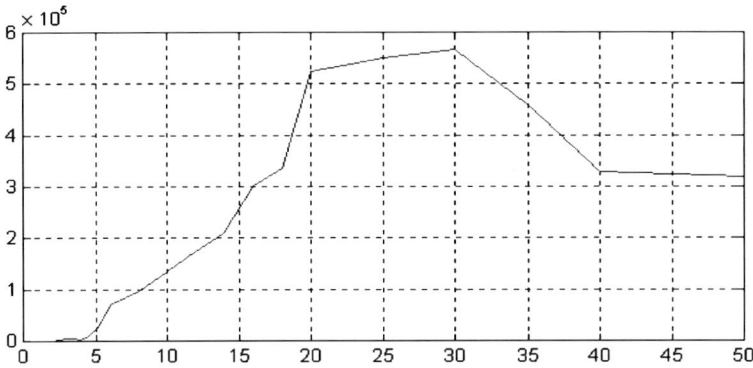

Fig. 10.6.20. Plot of N_* versus $\bar{u}$.

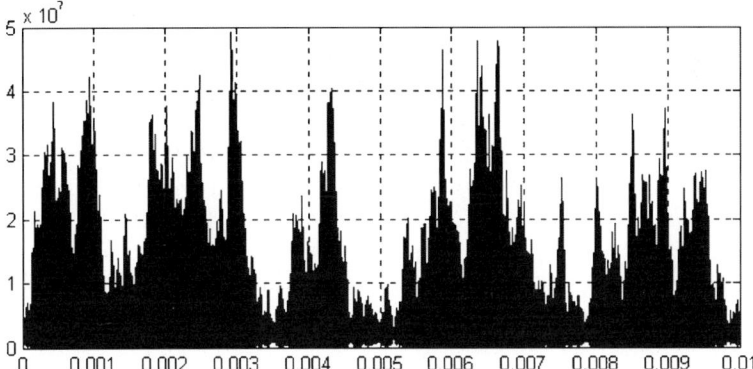

Fig. 10.6.21. Typical realization of process without control ($\bar{u} = 0$).

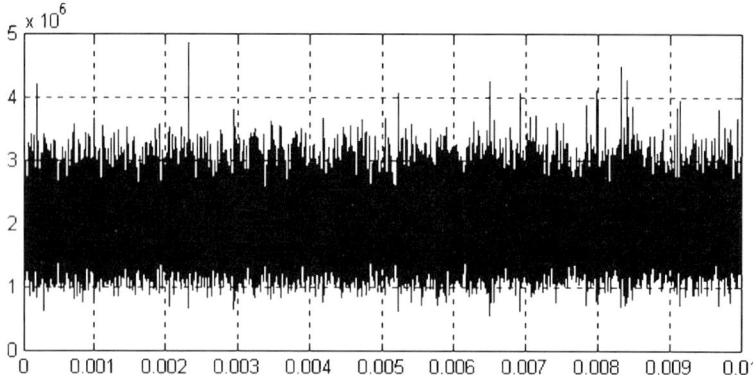

Fig. 10.6.22. Typical realization of process with control ($\bar{u} = 0.2 E_m$).

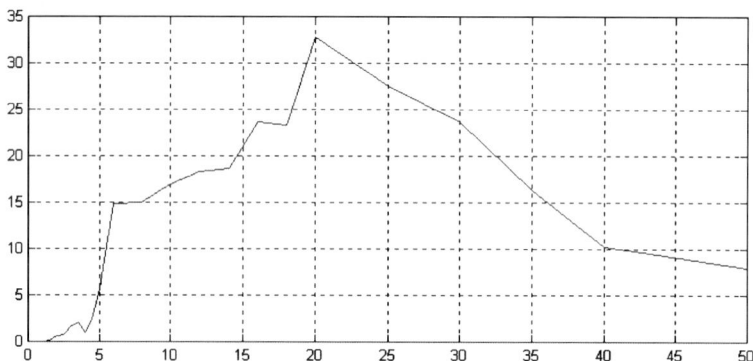

Fig. 10.6.23. Excitation index $\chi = N_*/\bar{u}$ versus $\bar{u}$.

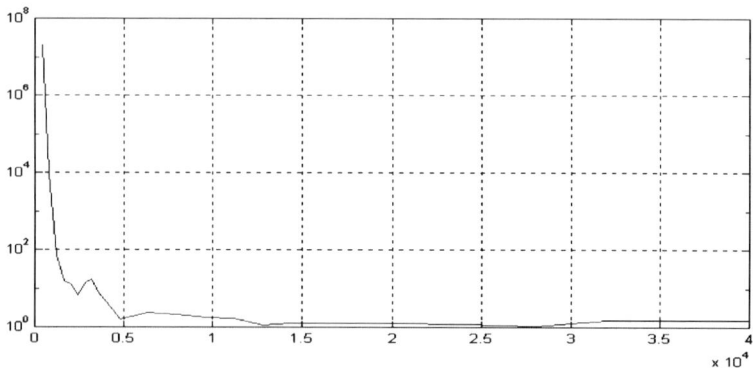

Fig. 10.6.24. Plot of ratio max $\{N_{\min}\}/\min\{N_{\min}\}$ versus $\bar{u}$.

10.6 Control of noise-induced transition

discharge stability (reliability). The minimum is taken over the time interval $0 \leq t \leq T$, where T is significantly larger than the typical time t_c.

Simulation results are shown in Fig. 10.6.20, where the scaling along the abscissa axis is chosen in percentage of the nominal voltage E_m. Typical time histories without of control ($\bar{u} = 0$) and with control ($\bar{u} = 0.2E_m$) are shown in Figs. 10.6.21 and 10.6.22. It is seen that the phase transition (the discharge failure) does not occur already for control amplitude greater than 5% of E_m. The further increase of the control amplitude $\bar{u} = 0$ leads to the increase of N_*. However, when $\bar{u} = 0$ reaches 20% of E_m, the growth is slowing down, and a further growth of N_* leads to the decrease of N_*. The effect of the sharp change of the current growth rate is better seen in Fig. 10.6.23 where the relative growth rate $\chi = N_*/\bar{u}$ normalized by the control amplitude is shown. The value of χ may be interpreted as the stochastic version of the excitability index introduced in Chapter 4 as a measure of ability of a system to absorb the energy from an external control. The ratio of the maximum to minimum of $N(t)$ over 10 trajectories (a measure of randomness of the effect) is shown in Fig. 10.6.24.

In the simulation of the system with the algorithm (10.6.33) it was assumed that the state of the system is available for measurement at each instant of numerical integration of the system (10.6.28,10.6.29), i.e., the system was considered as a continuous-time one. Since the integration step was chosen as $\Delta t = 2.5 \times 10^{-9}$ and time of one cycle of phase trajectory was about 2×10^{-6}, one cycle contained about 1000 measurements.

Implementation of frequent measurements in a real experiment may be rather involved. Therefore, it is interesting to study the possibility of more rare measurements. The properties of the system were examined for frequencies of measurements that are 20, 30, 40, and 50 times lower as compared to that used in getting the above data. The results are shown in the table and in Fig. 10.6.25.

Coefficient	$\min N(t) \times 10^5$	ratio
1:1	2.2471	1.6534
1:20	2.2004	1.7453
1:30	2.1668	1.8253
1:40	1.9828	1.8971
1:50	1.8822	1.9151

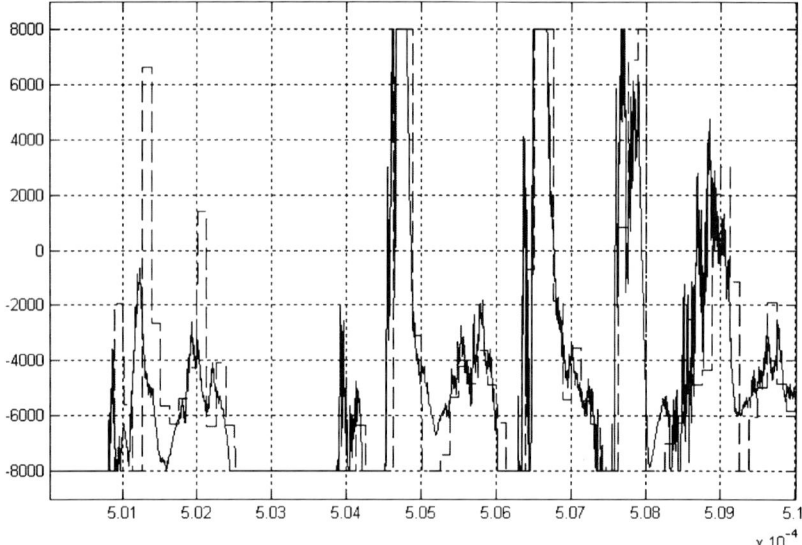

Fig. 10.6.25. Control signal plots for frequent (solid line) and rare (dashed line) measurements.

It is seen from these data that even 50 times less frequent measurements do not influence significantly the performance of the system: the minimum value of $N(t)$ decreases by 16%, while the measure of randomness $\max\{N_{\min}\}/\min\{N_{\min}\}$ increases only by 15%.

10.6.4 Discussion

According to the obtained results, while applying the suggested algorithm of the feedback control, one can keep the studied nonlinear system in a state where the tendency of the system to make a noise-induced transition to the dielectric state is suppressed. When introducing the control technique through variation in time of the voltage feeding the system, the pronounced positive result of the control has been revealed in computer simulations. It has been found that the effect can be observed at a rather low amplitude of the control action that, for the system parameters considered in the present work might be less than 5% of the nominal voltage feeding the device.

Similar control algorithm can be implemented to "correct" the dynamics of other systems which behavior can become undesirable due to the influence of stochastic forces. The suggested procedure can also be applied to control some model ecological systems with the intention to develop a technology of protecting the survival of threatened species. It might be of interest to extend the approach to the problems of controlling spatially extended systems which dynamics can produce spatiotemporal structures due to noise-induced resonance phenomena. As an experimental system which dynamics might be modeled, the "semiconductor – gas discharge gap" device could again be implemented. We add to the point that, being spatially extended, the device exemplifies a number of scenarios of self-organized behavior at some sets of experimental parameters [41]. Some patterns that arise in the device can be interpreted with a relatively simple theoretical model [42, 43] which is an extension of the lumped Astrov model (10.6.28, 10.6.29).

11

Conclusions: Looking into the Future

Modern physical studies are becoming more focused on the demands of practice. This means that a modern research project is a race for reaching a practical goal rather than a surfing over the ocean of knowledge. Another stimulus for pursuing practical goals is the grant-oriented system of funding; it is almost impossible nowadays to get funding for a research project driven by a curiosity. As a result, the concept of the goal is becoming more significant in physics.

In addition to direct problems (analysis) inverse problems (synthesis) also attract great interest. When solving synthesis or design problems an important auxiliary goal is to find a way to achieve a certain behavior of the system under investigation. Therefore, it is no surprise that the term "control" and the methods of modern control theory are becoming more frequent in physical journals. Such methods provide a number of tools for either analytically or numerically solving design problems for various dynamical systems. On the other hand, more applied studies stimulate a lot of basic research.

Systematic usage of the methods of modern control theory to study physical systems is a key feature of a new research area in physics that may be called *cybernetical physics*. The subject of cybernetical physics is focused on studying physical systems by means of feedback interactions with the environment. Its methodology heavily relies on the design methods developed in cybernetics. However, the approach of cybernetical physics differs from the conventional use of feedback in control applications (e.g., robotics, mechatronics) aimed mainly at driving a system to a prespecified position or a given trajectory. The cybernetical methodology may also be used to gain new insights into chemistry, biology, and environmental studies. Perhaps the future will provide new important contributions in this exciting field.

In this book some features of the control problems arising in physics were analyzed, and some general approaches to the solution of problems relating to energy control were presented. Energy transformation laws were established for two main classes of physical systems: conservative and dissipative. New problems demand for new notions (excitability index) and lead to the discovery of new phenomena (feedback resonance). Principles of the new area were illustrated by a number of examples showing efficiency of new approaches for the exploration and control of different phenomena in *macroworld* and *microworld*.

An attempt was made to show that nonlinear control design methods developed in control theory (cybernetics) may provide new interpretations and new insights for dynamical modeling and analysis of physical systems. Existence of strong analogies between dynamics of physical systems and control systems seems not very surprising – both are driven by similar variational principles.

Because of time and space constraints it was not possible to include a number of interesting and important applications of cybernetical methods in physics. For instance, the problem of controlled fusion – one of the biggest unsolved problems inherited by the 21st century from the 20th century, undoubtedly belongs to the area of cybernetical physics. Recent ground-breaking applications of control methods in other natural sciences (biology, chemistry), e.g., [415, 416] also need to be mentioned, as well as applications of other cybernetical methods: identification, pattern recognition, neural nets, and optimization.

There is a strong need for a lecture course on control methods for physicists. Although systematic textbooks are still missing, some surveys already exist. It is worth here to quote from a very readable recent survey [58]:

> Feedback and its big brother, control theory, are such important concepts that it is odd that they usually find no formal place in the education of physicists. On the practical side, experimentalists often need to use feedback. Almost any experiment is subject to the vagaries of environmental perturbations. Usually, one wants to vary a parameter of interest while holding all others constant. How to do this properly is the subject of control theory. More fundamentally, feedback is one of the great ideas developed (mostly) in the last century, with particularly deep consequences for biological systems, and all physicists should have some understanding of such a basic concept. (J. Bechhoefer. Feedback for physicists: A tutorial essay on control. *Reviews of Modern Physics*, V. 77, July 2005, 783–836.)

On the author's opinion, future course of cybernetical physics should cover, in addition to basic material and material of this book, such topics as control thermodynamics [67, 297, 410], control of particle beams [296], theory of dynamical materials [73, 406], laser control of processes in solids [302], control of stochastic systems [12, 50, 213, 277], studying universal structure and robustness properties of complex physical systems by means of control methods [63, 66, 97].

The horizons of a new interdisciplinary area look almost infinitely broad and interest in it is growing rapidly. Applied studies stimulate a lot of

11 Conclusions: Looking into the Future

basic research and cybernetical physics is an excellent example for this. At the present stage of its evolution it is important to increase the information exchange and dissemination of ideas between experts from different background and research areas. Evidence for this was provided by the success of the 1st International conference "Physics and Control" (Physcon 2003) that took place in August, 2003 in St. Petersburg, Russia with 250 experts from 32 countries participating. The four volumes of the conference proceedings contain more than 200 papers of total size about 1500 pages [358–361]. It was the opinion of many participants that it was necessary to organize another conference devoted to the same area. Then the 2nd International conference "Physics and Control" (Physcon 2005) took place in August, 2005, again in St. Petersburg, Russia. Information about the conference programs and Proceedings as well as further information about the area of cybernetical physics is available from the information website "Physics and Control Resources" at www.physcon.ru and the papers are available from the IEEE electronic library IEEE Xplore.

The author hopes that the publication of this book serves to the development of this new field and aid in its recognition by scientific communities.

References

1. Abarbanel, H.D.I., Korzinov, L., Mees, A.I., et al. Small force control of nonlinear systems to given orbits. *IEEE Trans. Circ. Syst. I* 1997, 44, 1018–1023.
2. Abarbanel, H., Rabinovich, M., Selverson, A., Bazhenov, M., Huerta, R., Sushchik, M., and Rubchinsky, L. Synchronization in neural assemblies. *Phys. Uspekhi* 1996, 39, 337–362.
3. Abed, E.H., and Fu, J.-H. Local feedback stabilization and bifurcation control. *Syst. Control Lett., Part I. Hopf Bifurcation* 1986, 7, 11–17; *Part II. Stationary Bifurcation* 1987, 8, 467–473.
4. Aero, E.L. Essentially nonlinear micromechanics of a medium with changeable periodic structure. *Adv. Mech.* 2002, 1(3), 131–176 (in Russian).
5. Aero, E.L., Andrievsky, B.R., Fradkov, A.L., and Vakulenko, S.A. Dynamics and nonlinear control of oscillations in a complex crystalline lattice. In: *Prepr. 16th IFAC World Congress Autom.Control*, Praha, July 2005.
6. Aero, E.L., Andrievsky, B.R., Fradkov, A.L., and Vakulenko, S.A. Dynamics and control of oscillations in a complex crystalline lattice. *Phys. Lett. A* 2006, 353(1), 24–29.
7. Afraimovich, V.S., Verichev, N.N., and Rabinovich, M.I. Synchronization of oscillations in dissipative systems. *Radiophysics and Quantum Electronics*, 1987, 795–803.
8. Akulenko, L.D. Parametric control of oscillations and rotations of physical pendulum (swings). *Appl. Math. Mech.* 1993, 57(2), 82–91.
9. Alekseev, V.V., and Loskutov, A. Destochastization of a system with a strange attractor by parametric interaction. *Moscow Univ. Phys. Bull.* 1985, 40(3), 46–49.
10. Alekseev, V.V., and Loskutov, A. Control of a system with a strange attractor through periodic parametric action. *Sov. Phys. Dokl.* 1987, 32, 1346–1348.
11. Alleyne, A. Reachability of chaotic dynamic systems. *Phys. Rev. Lett.* 1998, 80, 3751–3754.
12. Allison, A., Abbott, D. Control systems with stochastic feedback. *Chaos* 2001, 11(3), 715–724.
13. Alonso, A.A., and Ydstie, B.E. Stabilization of distributed systems using irreversible thermodynamics. *Automatica* 2001, 37(11), 1739–1755.
14. Alvarez-Gallegos, J. Nonlinear regulation of a Lorenz system by feedback linearization technique. *J. Dynamics Control* 1994, 4, 277–298.

15. Alvarez, J., Curiel, E., and Verduzco, F. Complex dynamics in classical control systems. *Syst. Control Lett.* 1997, 31, 277–285.
16. Amann, A., Schöll, E., Baba, N., and Just, W. Time delayed feedback control of chaos by spatio-temporal filtering. In: *Int. Conf. "Physics and Control,"* St. Petersburg, 2003 (http://www.rusycon.ru).
17. Amelkin, S.A., Martinás, K., and Tsirlin, A.M. Optimal control for irreversible processes in thermodynamics and microeconomic. *Autom. Remote Control* 2002, 63(4), 519–539.
18. Ananyevsky, M., Efimov, A., Fradkov, A., and Krivtsov, A. Resonance and speed-gradient design of control algorithms for dissociation of diatomic molecule ensembles. In: *Proc. Intern. Conf. "Physics and Control 2003,"* St. Petersburg, Aug. 20–22, 2003, pp. 867–878.
19. Ananyevskiy, M., Efimov, A., and Fradkov A. Control of quantum and classical molecular dynamics. In: *Prepr. 16th IFAC World Congress Autom.Control*, Praha, July 2005.
20. Anan'evskii, M.S., and Fradkov, A.L. Control of observables in the finite-level quantum systems. *Autom. Remote Control* 2005, 66(5), 734–745.
21. Ananjevsky, M.S., Vetchinkin, A.S., Sarkisov, O.M., Umanskii, S.Ya., Fradkov, A.L., and Zotov, Yu.A. Quantum control of dissociation of an iodine molecule by one and two femtosecond laser pulses excitation. In: *Proc. 2nd Intern.Conf. "Physics and Control,"* IEEE, St. Petersburg, 2005, pp. 636–641.
22. Andresen, B. *Finite-time Thermodynamics*. Copenhagen: University of Copenhagen, 1983.
23. Andrievsky, B.R. Adaptive synchronization methods for signal transmission on chaotic carriers. *Math. Comput. Simulation* 2002, 58(4–6), 285–293.
24. Andrievsky, B.R., Guzenko, P.Yu., and Fradkov, A.L. Control of nonlinear oscillations of mechanical systems by speed-gradient method. *Autom. Remote Control* 1996, 57(4), 456–467.
25. Andrievsky, B.R., and Fradkov, A.L. *Selected Chapters of Control Theory with Examples in MATLAB*. Nauka, St. Petersburg, 1999 (in Russian) 467.
26. Andrievsky, B.R., and Fradkov, A.L. Feedback resonance in single and coupled 1-DOF oscillators. *Intern. J. Bifurcations Chaos* 1999, 10, 2047–2058.
27. Andrievsky, B.R., and Fradkov, A.L. Control of chaos. I. Methods. *Autom. Remote Control* 2003, 64(5), 673–713.
28. Andrievsky, B.R., and Fradkov, A.L. Control of Chaos. II. Applications. *Autom. Remote Control* 2004, 65(4), 505–533.
29. Andrievsky, B.R., Stotsky, A.A., and Fradkov A.L. Velocity gradient algorithms in control and adaptation problems. A survey. *Autom. Remote Control* 1998, 12, 1533–1563.
30. Andronov, A.A., and Witt, A. Towards mathematical theory of capture. *Arch. Electrotechnik* 1930, 24(1), 99–110.
31. Andronov, A., Vitt, A., and Khaikin, S. *Theory of Oscillations*. Pergamon Press, Oxford (in Russian: 1st edn., 1937; 2nd edn. 1959).
32. Anishchenko, V.S., Neiman, A.B., Moss, F., and Schimansky-Geier, L. Stochastic resonance: Noise-enhanced order. *Phys. Uspekhi* 1999, 169(1), 7–37.
33. Anishchenko, V.S., Astakhov, V.V., Neiman, A.B., Vadivasova, T.E., and Schimansky-Geier, L. *Nonlinear Dynamics of Chaotic and Stochastic Systems*. Springer Berlin-Heidelberg, 2002.
34. Appleton, E. The automatic synchronization of triode oscillators. *Proc. Cambridge Philos. Soc.* 1922, 21(3), 231–248.

35. Aranson, I., Meerson, B., and Tajima, T. Excitation of solitons by an external resonant wave with a slowly varying phase velocity. *Phys. Rev. A* 1992, 45, 7500.
36. Arecchi, F.T., and Boccaletti, S. Adaptive strategies for recognition, noise filtering, control, synchronization and targeting of chaos. *Chaos* 1997, 7, 621–634.
37. Arnol'd, V.I. *Mathematical Methods of Classical Mechanics*. Springer-Verlag, Berlin, 1978 (in Russian: 1974).
38. Assion, A., Baumert, T., Bergt, M., Brixner, T., Kiefer, B., Seyfried, V., Strehle, M., and Gerber, G. Control of chemical reactions by feedback-optimized phase-shaped femtosecond laser pulses. *Science* 1998, 282, 919.
39. Astrov Yu.A. Dynamic Properties of Discharge Glow in a Device with Resistive Electrode. A. F. Ioffe Physico-Technical Institute Report No. 1255, 1988 (in Russian).
40. Astrov, Yu.A., Portsel, L.M., Teperick, S.P., Willebrand, H., and Purwins, H.-G. Speed properties of a semiconductor – discharge gap IR image converter studied with a streak camera system *J. Appl. Phys.* 1993, 74, 2159–2166.
41. Astrov, Yu.A., Ammelt, E., Teperick, S.P., and Purwins, H.-G. Hexagon and stripe Turing structures in a gas discharge system. *Phys. Lett. A* 1996, 211, 184–190.
42. Astrov, Yu.A., and Logvin, Yu.A. Formation of clusters of localized states in a gas discharge system via a self-completion scenario. *Phys. Rev. Lett.* 1997, 79, 2983–2986.
43. Astrov, Yu.A. Phase transition in an ensemble of dissipative solitons of a Turing system. *Phys. Rev. E* 2003, 67, 035203-1–035203-4.
44. Astrov, Yu.A., Fradkov, A.L., and Guzenko, P.Yu. Suppression of a noise-induced transition by feedback control. In: *Proc. 2nd Intern. Conf. "Physics and Control,"* IEEE, St. Petersburg, 2005, pp. 662–667.
45. Åström, K.J., and Furuta, K. Swinging up a pendulum by energy control. *Automatica* 2000, 36(2), 287–295.
46. Belavkin V.P. Optimal Measurement and Control in Quantum Dynamical Systems. Preprint Instytut Fizyki, Copernicus University, Torun', 1979, 411, 3–38 (see also ArXiv.org, quant-ph/0208108).
47. Babitsky, V.I., and Chitaev, M.I. Adaptive high-speed resonant robot. *Mechatronics* 1996, 6, 897–913.
48. Babitsky, V.I., and Krupenin, V.L. *Vibration of Strongly Nonlinear Discontinuous Systems*. Springer Berlin-Heidelberg, 2001 (in Russian: 1985).
49. Babloyantz, A., Krishchenko, A.P., and Nosov, A. Analysis and stabilization of nonlinear chaotic systems. *Comput. Math. Appl.* 1997, 34, 355–368.
50. Balanov, A.G., Janson, N.B., and Schöll, E. Control of noise-induced oscillations by delayed feedback. *Physica D* 2004, 199, 1–12.
51. Baillieul, J., Brockett, R.W., and Washburn, R.B. Chaotic motion in nonlinear feedback systems. *IEEE Trans. Circ. Syst. I* 1980, 27, 990–997.
52. Bak, P.E., and Yoshino, R. A Dynamical Model of Maxwell's Demon and confinement systems. *Contrib. Plasma Phys.* 2000, 40(3–4), 227–232.
53. Bandrauk A.D., Fujimura, Y., and Gordon, R.J. (eds.) *Laser Control and Manipulation of Molecules*. Oxford University Press, Oxford, 2002.
54. Bardeen, C.J., Yakovlev, V.V., Wilson, K.R., Carpenter, S.D., Weber P.M., and Warren, W.S. Feedback quantum control of molecular electronic population transfer. *Chem. Phys. Lett.* 1997, 280, 151.

55. Basso, M., Genesio, R., and Tesi, A. Stabilizing periodic orbits of forced systems via generalized Pyragas controllers. *IEEE Trans. Circ. Syst. I* 1997, 44, 1023–1027.
56. Basso, M., Genesio, R., Giovanardi, L. et al. On optimal stabilization of periodic orbits via time delayed feedback control. *Int. J. Bifurcation Chaos* 1998, 8, 1699–1706.
57. Basso, M., Genesio, R., Giovanardi, L., and Tesi A. Frequency domain methods for chaos control. In: *Controlling Chaos and Bifurcations in Engineering Systems*. G. Chen (ed.) CRC Press, Boca-Raton, FL 1999, pp. 179–204.
58. Bechhoefer, J. Feedback for physicists: A tutorial essay on control. *Rev. Mod. Phys.* 2005, 77, 783–836.
59. Bejan, A. "Entropy generation minimization: The new thermodynamics of finite-size devices and finite-time processes. *J. Appl. Phys.* 1996, 79, 1191–1218.
60. Beletskii, V.V. *Two-Legged Walking*. Nauka, Moscow, 1984 (in Russian).
61. Beletskii, V.V. Nonlinear effects in dynamics of controlled two-legged walking. In: *Nonlinear Dynamics in Engineering Systems*. Springer Berlin-Heidelberg, 1990, pp. 17–26.
62. Bellman, R., Bentsman, J., and Meerkov, S. Vibrational control of nonlinear systems. *IEEE Trans. Autom. Control* 1986, AC-31(8), 710–724.
63. Belykh, V.N., Belykh, I.V., Hasler, M. Hierarchy and stability of partially synchronous oscillations of diffusively coupled dynamical systems. *Phys. Rev. E* 2000, 62 (5), 6332–6345.
64. Belykh, V.N., Belykh, I.V., and Mosekilde, E. Cluster synchronization modes in an ensemble of coupled chaotic oscillators. *Phys. Rev. E* 2001, 63, 036216.
65. Belykh, V.N., Belykh, I.V., Hasler, M., and Nevidin, K.V. Cluster synchronization in three-dimensional lattices of diffusively coupled oscillators. *Int. J. Bifurcation Chaos* 2003, 13(4), 755–779.
66. Belykh, V.N., Osipov, G.V., Kucklander, N., Blasius, B., and Kurths, J. Automatic control of phase synchronization in coupled complex oscillators. *Physica D* 2005, 200, 81–104.
67. Berry, R.S., Kazakov, V.A., Sieniutycz, S., Szwast, Z., and Tsirlin, A.M., *Thermodynamic Optimization of Finite Time Processes*. Wiley, New York, 2000.
68. Beta, C., and Mikhailov, A.S. Controlling spatiotemporal chaos in oscillatory reaction-diffusion systems by time-delay autosynchronization. *Physica D* 199, 2004, 173–184.
69. Bishop, S.R., Xu, D.L., and Clifford, M.J. Flexible control of parametrically excited pendulum. *Proc. R. Soc. Lond. A* 1996, 452, 1789–1806.
70. Bleich, M.E., Hochheiser, D., Moloney, J.V., and Socolar, J.E.S. Controlling extended systems with spatially filtered, time-delayed feedback. *Phys. Rev. E* 1997, 55, 2119–2126.
71. Blekhman, I.I. *Synchronization of Dynamical Systems*. Nauka, Moscow, 1971, 894 (in Russian).
72. Blekhman, I.I. *Synchronization in Science and Technology*. ASME Press, New York, 1988 (in Russian: 1981).
73. Blekhman, I. *Vibrational Mechanics*. World Scientific, Singapore, 2000 (in Russian: 1994).
74. Blekhman, I.I., and Fradkov, A.L. On the general definitions of synchronization. In: *Selected Topics in Vibrational Mechanics*. I.I. Blekhman (ed.). World Scientific, Singapore, 2004, pp. 179–188.

75. Blekhman, I.I., Fradkov, A.L., Nijmeijer, H., and Pogromsky, A.Yu. On self-synchronization and controlled synchronization. *Syst. Control Lett.* 1997, 31, 299–305.
76. Blekhman, I.I., Fradkov, A.L., Tomchina, O.P., and Bogdanov, D.E. Self-synchronization and controlled synchronization: General definition and example design. *Math. Comput. Simulation* 2002, 58(4–6), 367–384.
77. Bliman, P.-A., Krasnosel'skii, A.M., Sorine, M., and Vladimirov, A.A. Nonlinear resonance in systems with hysteresis. *Nonlinear Analysis: Theory Appl.* 1996, 27(5), 561–577.
78. Boccaletti, S., Maza, D., Mancini., H., Genesio., R., and Arecchi, F.T. Control of defects and spacelike structures in delayed dynamical systems. *Phys. Rev. Lett.* 1997, 79, 5246–5249.
79. Boccaletti, S., Bragard, J., and Arecchi, F.T. Controlling and synchronizing space time chaos. *Phys. Rev. E* 1999, 59, 6574–6578.
80. Boccaletti, S., Grebogi, C., Lai, Y.C. et al. The control of chaos: theory and applications. *Phys. Rep.* 2000, 329, 103–197.
81. Boccaletti, S., Kurths, J., Osipov, G., Valladares, D.L., and Zhou, C.S. The synchronization of chaotic systems. *Phys. Rep.* 2002, 366, 1–101.
82. Boccaletti, S., Pecora, L., and Pelaez, A. Unifying framework for synchronization of coupled systems. *Phys. Rev. E* 2001, 63, 066219.
83. Bogoliubov, N.N., Mitropolski, Yu.A., *Asymptotic Methods in the Theory of Nonlinear Oscillations.* Gordon-Breach, New York, 1961.
84. Bollt, E.M. Controlling chaos and the inverse Frobenius–Perron problem: Global stabilization of arbitrary invariant measures. *Int. J. Bifurcation Chaos.* 2000, 10(5), 1033–1050.
85. Bondarko, V.A. and Yakubovich, V.A. The method of recursive goal inequalities in adaptive control theory. *Int. J. Adaptive Control Sign. Proc.* 1992, 6(3), 141–160.
86. Born, M., and Huang, K. *Dynamic Theory of Crystalline Lattices.* Oxford University Press, Oxford, 1998.
87. Brandt, M.E., Shih, H.T., and Chen, G.R. Linear time-delay feedback control of a pathological rhythm in a cardiac conduction model. *Phys. Rev. E* 1997, 56, R1334–R1337.
88. Brillouin, L. *Science and Information Theory*, 2nd edn. Academic Press, New York, 1962.
89. Brockett, R.W. Control theory and analytical mechanics. In: *Geometric Control Theory, Lie Groups. V. VII.* Martin, C., and Hermann R. (eds.). *Mat. Sci.* Press, Brookine, MA, 1977, 1–48.
90. Brown, R., and Kocarev, L. A unifying definition of synchronization for dynamical systems. *Chaos* 2000, 10(2), 344–349.
91. Brown, E., and Rabitz, H. Some mathematical and algorithmic problems of control of quantum dynamics. *J. Math. Chem.* 2002, 31(1), 17–63.
92. Brumer, P.W., and Shapiro, M. *Principles of the Quantum Control of Molecular Processes.* Wiley-Interscience, Hoboken NJ, 2003.
93. Butkovskii, A.G. *Distributed Control Systems.* American Elsevier, New York, 1969 (in Russian: 1965).
94. Butkovskii, A.G., and Samoilenko, Yu.I. Control of quantum-mechanical systems, I, II. *Autom. Remote Control* 1979, 40, 4, 5.

95. Butkovskii, A.G., and Samoilenko, Yu.I. *Control of quantum-mechanical systems and processes.* Kluwer Academic Publishers, Dordrecht, 1990 (in Russian: 1984).
96. Byrnes, C., Isidori, A., and Willems, I.C. Passivity, feedback equivalence and the global stabilization of minimum phase nonlinear systems. *IEEE Trans. Autom. Control* 1991, AC-36(11), 1228–1240.
97. Carlson, J.M., Doyle, J. Highly optimized tolerance: A mechanism for power laws in designed systems. *Phys. Rev. E* 1999, 60, 1412–1427.
98. Chacón R. Maintenance and suppression of chaos by weak harmonic perturbations: A unified view *Phys. Rev. E* 2001, 9, 1737–1430.
99. Chacón, R. *Control of Homoclinic Chaos by Weak Periodic Perturbations.* World Scientific Series on Nonlinear Science. Ser. A. World Scientific, Singapore, 2002.
100. Chang, K., Kodogeorgiou, A., Hubler, A., and Jackson, E.A. General resonance spectroscopy. *Physica D* 1991, 51, 99–108.
101. *Chaos: An Interdisciplinary Journal of Nonlinear Science.* Focus issue: Control and synchronization of chaos. Eds. Ditto, W.L., and Showalter K., 1997, 7(4).
102. *Chaos: An Interdisciplinary Journal of Nonlinear Science.* Focus issue: Control and synchronization in chaotic dynamical systems. J. Kurths, S. Boccaletti, C., Grebogi, and Y.-C. Lai (eds.) 2003, 13(1), 1–419.
103. Chen, G., and Dong, X. *From Chaos to Order: Perspectives, Methodologies and Applications.* World Scientific, Singapore, 1998.
104. Chen, G.R., and Lai, D.J. Feedback anticontrol of discrete chaos. *Int. J. Bifurcation Chaos* 1998, 8, 1585–1590.
105. Chen, L.Q., and Liu, Y.Z. A modified exact linearization control for chaotic oscillators. *Nonlinear Dynamics* 1999, 20, 309–317.
106. Chernousko, F.L. Some problems of optimal control with a small parameter. *J. Appl. Math. Mech.* 1968, (3), 12–22.
107. Chung, C.C., and Hauser, J. Nonlinear control of a swinging pendulum. *Automatica* 1995, 31(6), 851–862.
108. Churilova, M.Y. Conditions for stabilization of nonlinear systems by harmonic external action. *Autom Remote Control* 1994, (3).
109. Ciofini, M., Labate, A., Meucci, R., and Arecchi, F.T. Experimental control of chaos in a delayed high-dimensional system. *Europ. Phys. J. D* 1999, 7, 5–9.
110. Codreanu, S., and Danca, M. Suppression of chaos in a one–dimensional mapping. *J. Biol. Phys.* 1997, 23, 1–9.
111. *Control of Mechatronic Vibrational Units.* Blekhman, I.I., and Fradkov A.L. (eds.). Nauka, St. Petersburg, 2001, 278.
112. Cuomo, K.M., Oppenheim, A.V., and Strogatz, S.H. Synchronization of Lorenz-based chaotic circuits with application to communications. *IEEE Trans. Circ. Syst. II.* 1993, 40(10), 626–633.
113. Curzon, F.L., and Ahlburn, B. Efficiency of a Carnot engine at maximum power output. *Am. J. Phys.* 43, 22–24, 1975.
114. Dahleh, M., Pierce, A., Rabitz, H., and Ramakrishna , V. Control of molecular motion. *Proc. IEEE* 1996, 84(1), 7–15.
115. D'Alessandro, D., Dahleh, M., and Mezic, I. Control of mixing in fluid flow: A maximum entropy approach. *IEEE Trans. Autom. Control* 1999, 44(10), 1852–1863.

116. Dedieu, H., Kennedy, M.P., and Hasler, M. Chaos shift keying: Modulation and demodulation of chaotic carrier using self-synchronized Chua's circuits. *IEEE Trans. Circ. Syst.* II: Analog Digital Signal Processing 1993, 40(10), 634–642.
117. Demidovich, B.P. *Lectures on Mathematical Stability Theory.* Nauka, Moscow, 1967.
118. Devaney, R. *A First Course in Chaotic Dynamical Systems.* Addison-Wesley, New York, 1992.
119. Diem, M. *Introduction to Modern Vibrational Spectroscopy.* Wiley, Hoboken, NJ, 1993.
120. Ditto, W.L., Rauseo, S.N., and Spano, M.L., Experimental control of chaos. *Phys. Rev. Lett.* 1990, 65, 3211–3214.
121. Ditto, W.L., Spano, M.L., In, V., Neff, J., Meadows, B., Langberg, J.J., Bolmann, A., and McTeague, K. Control of human atrial fibrillation. *Int. J. Bifurcation Chaos* 2000, 10, 593–601.
122. DiVincenzo, D.P. Quantum computation. *Science* 1996, 270, 255–261.
123. Dudnik, E.N., Kuznetsov, Yu.I., Minakova, I.I., and Romanovskii, Yu.M. Synchronization in systems with strange attractors. *Moscow Univ. Phys. Bull. Ser.3* 1983, 24(4), 84–87.
124. Dykstra, R., and Tang, D.Y., Heckenberg, N.R. Experimental control of single-mode laser chaos by using continuous, time-delayed feedback. *Phys. Rev. E* 1998, 57, 6596–6598.
125. Einstein, A. *Out of My Later Years.* Thames and Hudson, New York, 1950, 114.
126. Efimov, A.A., Fradkov, A.L., and Krivtsov, A.M. Feedback design of control algorithms for dissociation of diatomic molecules. In: *Proc. Europ. Control Conf.* Cambridge, UK, 1–4 Sept. 2003.
127. Enikov, E., and Stepan, G. Microchaotic motion of digitally controlled machines. *J. Vibration Control* 1998, 4, 427–443.
128. Escalona, J., and Parmananda, P. Noise-aided control of chaotic dynamics in a logistic map. *Phys. Rev. E* 2000, 61, 5987–5989.
129. Farkas, H., and Noszticzius, Z., On the non-linear generalization of the Gyarmati principle and theorem. *Annalen der Physik* 27, 1971, 341–348.
130. Fajans, J., and Friedland, L. Autoresonant – nonstationary – excitation of pendulums, Plutinos, plasmas, and other nonlinear oscillators. *Am. J. Phys.* 2001, 69(10), 1096.
131. Flugge, S. *Practical Quantum Mechanics 2*, Springer, Berlin, 1971.
132. Fomin, V., Fradkov, A., and Yakubovich, A. *Adaptive Control of Dynamical Systems.* Nauka, Moscow, 1981 (in Russian).
133. Fouladi, A., and Valdivia, J.A. Period control of chaotic systems by optimization. *Phys. Rev. E* 1997, 55, 1315–1320.
134. Fradkov, A.L. Speed-gradient scheme and its applications in adaptive control. *Autom. Remote Control* 1979, 40(9), 1333–1342.
135. Fradkov, A.L. *Adaptive Control in Complex Systems.* Nauka, Moscow, 1990 (in Russian).
136. Fradkov, A.L. Speed-gradient laws of control and evolution. In: *Prepr. 1st European Control Conf.*, Grenoble, 1991, 1865–1870.
137. Fradkov, A.L. Nonlinear adaptive control: regulation, tracking, oscillations. In: *Proc. 1st IFAC Workshop* "New trends in design of control systems." Smolenice, 1994, 426–431.

138. Fradkov, A.L. Swinging control of nonlinear oscillations. *Intern. J. Control* 1996, 64(6), 1189–1202.
139. Fradkov, A.L. Exploring nonlinearity by feedback. Memorandum No. 1447, Faculty of Mathematical Sciences, University of Twente, June 1998.
140. Fradkov, A.L. Exploring nonlinearity by feedback. *Physica D* 1999, 128(2–4), 159–168.
141. Fradkov, A.L. Investigation of physical systems by means of feedback. *Autom. Remote Control* 1999, 60(3), 3–22.
142. Fradkov, A.L. Feedback resonance in nonlinear oscillators. In: *Proc. 5th Europ. Contr. Conf.*, Karlsruhe, 1999.
143. Fradkov, A.L. A nonlinear philosophy for nonlinear systems. In: *Proc. 39th IEEE Conf. Decisions and Control.* Sydney, 12–15 Dec. 2000, 4397–4402.
144. Fradkov, A.L., Physics and control: Exploring physical systems by feedback. In: *Prepr. 5th IFAC Symp.* "Nonlinear Control Systems" (NOLCOS'01). St.Petersburg, 4–6 July 2001, 1503–1509.
145. Fradkov, A.L. Control of chaotic systems. In: *Encyclopedia of Life Support Systems (EOLSS)*, Developed under the auspices of the UNESCO. Eolss Publishers, Oxford, UK, 2003 [http://www.eolss.net].
146. Fradkov, A.L. Speed-gradient approach to modeling dynamics of physical systems. In: *Proc. Europ. Control Conf. (ECC'03)*. Cambridge, UK, 1–4 Sept. 2003.
147. Fradkov, A.L. *Cybernetical Physics: Principles and Examples.* Nauka, St.Petersburg, 2003, p. 208 (in Russian).
148. Fradkov, A.L. Application of cybernetical methods in physics. *Physics-Uspekhi* 2005, 48(2), 103–127.
149. Fradkov, A.L., Ananyevsky, M.S., and Efimov, A.A. Cybernetical physics and control of molecular systems. "Frontiers of Nonlinear Physics". In: *Proc. II Intern. Conf.*, Nizhny Novgorod, 2004.
150. Fradkov, A.L., and Guzenko, P.Yu. Adaptive control of oscillatory and chaotic systems based on linearization of Poincaré map. In: *Proc. 4th Europ. Control Conf.*, Brussels, 1997.
151. Fradkov, A.L., Guzenko, P.Yu., Kukushkin, S.A., and Osipov, A.V. Dynamics and control of thin film growth from multicomponent gas. *J. Phys. D: Appl. Phys.* 1997, 30(20), 2794–2797.
152. Fradkov, A., Guzenko, P., and Pavlov, A. Adaptive control of recurrent trajectories based on linearization of Poincaré map. *Int. J. Bifurcation Chaos* 2000, 10, 621–637.
153. Fradkov, A.L., and Evans, R.J. Control of chaos: Methods and applications in engineering. *Annu. Rev. Control* 2005, 29(1), 33–56.
154. Fradkov, A.L., Krivtsov, A.M., and Efimov, A.A. Dissociation of diatomic molecule by speed-gradient feedback control. *Differential Equations Control Processes* 2001 (4), 36–46, http://www.neva.ru/journal/.
155. Fradkov, A.L., Krivtsov, A.M., and Efimov, A.A. http://ArXiv.org. *E-print physics/0201020*, Jan 10, 2002.
156. Fradkov, A.L., Leonov, G.A., and Nijmeijer, H. Frequency-domain conditions for global synchronization of nonlinear systems driven by a multiharmonic external signal. In: *Proc. 5th European Control Conf.*, Karlsruhe, 1999.
157. Fradkov, A.l., Miroshnik, I.V., and Nikiforov, V.O. *Nonlinear and Adaptive Control of Complex Systems.* Kluwer Academic Publishers, Dordrecht, 1999.

158. Fradkov, A.L., Makarov, I.A., Shiriaev, A.S., and Tomchina, O.P. Control of oscillations in Hamiltonian systems. In: *4th European Control Conf.* (ECC'97). Brussels, 1997.
159. Fradkov, A.L., and Markov, A.Yu. Adaptive synchronization of chaotic systems based on speed gradient method and passification. *IEEE Trans. Circ. Syst. Part I* 1997 (10), 905–912.
160. Fradkov, A.L., Nijmeijer, H., and Markov, A.Yu. Adaptive observer-based synchronization for communication. *Int. J. Bifurcations Chaos* 2000, 10(12), 2807–2813.
161. Fradkov, A.L., Nikiforov, V.O., and Andrievsky, B.R. Adaptive observers for nonlinear nonpassifiable systems with application to signal transmission. In: *Proc. 41th IEEE Conf. Dec. Control.* Las Vegas, 2002, 4706–4711.
162. Fradkov, A.L., and Pogromsky, A.Yu. Methods of adaptive control of chaotic systems. In: *Proc. of 12th IFAC World Congress.* San Francisco, 1996. Vol. K, pp. 185–190.
163. Fradkov, A.L., and Pogromsky, A.Yu. Speed gradient control of chaotic continuous-time systems. *IEEE Trans. Circ. Syst. I* 1996, 43, 907–913.
164. Fradkov A.L., and Pogromsky A.Yu. *Introduction to Control of Oscillations and Chaos.* World Scientific, Singapore, 1998.
165. Fradkov, A.L., and Shiegin, V.V. Stabilization of the energy of oscillations with application to a pendulum with controlled suspension point. *J. Comput. Syst. Sci. Intern.* 1999 (2), 179–184.
166. Friedel, H., Grauer, R., and Spatschek, K.H. Controlling chaotic states of a Pierce diode. *Phys. Plasmas* 1998, 5, 3187–3194.
167. Frieden, B.R., and Soffer, B.H. Lagrangians of physics and the game of Fisher information transfer. *Phys. Rev. E* 1995, 52, 227486.
168. Frieden, B.R. *Physics from Fisher Information.* Cambridge University Press, Cambridge, 1998.
169. Frieden, B.R., and Soffer, B.H. A critical comparison of three information-based approaches to physics. *Foundations Phys. Lett.* 2000, 13(1), 89–96.
170. Fujisaka, H., and Yamada, T. Stability theory of synchronized motion in coupled-oscillator systems. *Prog. Theor. Phys.* 69, 1983, 32–47.
171. Gad-el-Hak, M. *Flow Control: Passive, Active, and Reactive Flow Management.* Cambridge University Press, London, 2000.
172. Gade, P.M. Feedback control in coupled map lattices. *Phys. Rev. E* 1998, 57, 7309–7312.
173. Garfinkel, A., Spano, M.L., Ditto, W.L., and Weiss, J.N. Controlling cardiac chaos. *Science* 1992, 257, 1230–1235.
174. Gelig, A.Kh. Stabilization of impulse systems by periodic external action. *Autom. Remote Control* 1990 (4).
175. Genesio, R., and Tesi, A. Harmonic balance methods for the analysis of chaotic dynamics in nonlinear systems. *Automatica* 1992, 28(3), 531–548.
176. Gershenfeld, N. *The Physics of Information Technology.* Cambridge University Press, Cambridge, 2000.
177. Glad, T., and Ljung, L. *Control Theory.* Taylor & Francis, London and New York, 2000.
178. Glansdorff, P., and Prigogine, I. *Thermodynamics of Structure, Stability and Fluctuations.* Wiley, New York, 1971.
179. Glorieux, P. Control of chaos in lasers by feedback and nonfeedback methods. *Int. J. Bifurcation Chaos* 1998, 8, 1749–1758.

180. Goggin, M.E., and Milonni, P.W. Driven Morse oscillator: Classical chaos, quantum theory and photodissociation. *Phys. Rev. A* 1988, 37(3), 796.
181. Goggin, M.E., and Milonni, P.W. Driven Morse oscillator: Classical chaos and quantum theory for two-frequency dissociation. *Phys. Rev. A* 1988, 38(10), 5174.
182. Goldobin, D.S., and Pikovsky, A.S. Synchronization and desynchronization of self-sustained oscillators by common noise. *Phys. Rev. E* 2005, 71, 045201.
183. Gora, P., and Boyarsky, A. A new approach to controlling chaotic systems. *Physica D* 1998, 111(1–4), 1–15.
184. Grebogi, C., Ott, E., and Yorke, J. *Phys. Rev. Lett.* 1986, 57, 1284–1287.
185. Grigoriev, R.O., Cross, M.C., and Schuster, H.G. Pinning control of spatiotemporal chaos. *Phys. Rev. Lett.* 1997, 79, 2795–2798.
186. Grivopoulos. S., and Bamieh, B. Lyapunov-based control of quantum systems. In: *Proc. 42nd IEEE Conf. Decision and Control.* 2003, pp. 434–438.
187. Grivopoulos, S. and Bamieh, B. Optimal Population Transfers for a Quantum System in the Limit of Large Transfer Time. *Proc. American Control Conf.*, 2004, pp. 2481–2486.
188. Guldberg, A., and Billing, G.D. Laser-induced dissociation of hydrogen fluoride. *Chem. Phys. Lett.* 1991, 186(2–3), 229.
189. Guckenheimer, J., and Holmes, P. *Nonlinear Oscillations, Dynamical Systems and Bifurcations of Vector Fields.* Springer-Verlag, Berlin, 1983.
190. Gurtovnik, A.S., and Neimark, Yu.I. On synchronization of dynamical systems. *Appl. Math. Mech.* 1974 (5).
191. Guzenko, P.Yu., and Fradkov, A.L. Gradient control of Henon map dynamics. *Int. J. Bifurcations Chaos* 1997, 7(3), 701–705.
192. Gyarmati, I. *Non-equilibrium Thermodynamics. Field Theory and Variational Principles.* Springer-Verlag, Berlin, 1970.
193. Gzil, M. *The Method of Maximum Entropy.* World Scientific, Singapore, 1995.
194. Haken, H. *Information and Self-Organization.* Springer, Berlin Heidelberg, 2000.
195. Hangos, K.M., Bokor, J., and Szederkenyia, G. Hamiltonian view on process systems. *AIChE J.* 2001, 47(8), 1819–1831.
196. Hartley, R.V.L. Transmission of Information. *Bell Syst. Technical J.* July 7, 1928, 535.
197. Helmke, U., Pratzelwolters, D., and Schmid, S. Adaptive synchronization of interconnected linear systems. *IMA J. Math. Control.* 1991, 8(4), 397–408.
198. Hohmann, W., Kraus, M., and Schneider, F.W. Recognition in excitable chemical reactor networks. Experiments and model-simulations. *J. Phys. Chem. A* 1997, 101, 7364–7370.
199. Horsthemke W., and Lefever R. *Noise-Induced Transitions. Theory and Applications in Physics, Chemistry, and Biology.* Springer-Verlag, Berlin, 1984.
200. Hu, G., Qu, Z., and He, K. Feedback control of chaos in spatiotemporal systems. *Int. J. Bifurcation Chaos* 1995, 5, 901–936.
201. Hu, G., Xiao, J.H., Gao, J.H., Li, X.M., Yao, Y.G., and Hu, B.B. Analytical study of spatiotemporal chaos control by applying local injections. *Phys. Rev. E* 2000, 62, R3043–R3046.
202. Huang, G.M., Tarn, T.J., and Clark, J.W. On the controllability of quantum-mechanical systems. *J. Math. Phys.* 1983, 24, 2608.
203. Hubler, A., and Lusher, E. Resonant stimulation and control of nonlinear oscillators. *Naturwissenschaft* 1989, 76, 67–72.

204. Hunt, E.R. Stabilizing high-period orbits in a chaotic system – the diode resonator. *Phys. Rev. Lett.* 1991, 67, 1953–1955.
205. Ibragimov, Kh.O., Aliev, K.M., Kamilov, I.K., and Abakarova, N.S. Chaos in germanium oscillistor. In: *2003 Int. Conf. "Physics and Control" Proc.* Fradkov, A.L., and Churilov, A.N. (eds.) St.Petersburg, Russia, 2003, 680–682.
206. *IEEE Trans. on Circuits and Systems.* Special issue on applications of chaos in modern communication systems. L. Kocarev, G.M. Maggio, M., Ogorzalek, L. Pecora, and K. Yao. (eds.). 2001, 48(12)
207. *IEEE Trans. on Circuits and Systems.* Special issue: Chaos control and synchronization. M. Kennedy, and M. Ogorzalek (eds.). 1997, 44(10).
208. *International Journal of Bifurcations and Chaos.* Theme issue: Control and Synchronization of Chaos. Chen G., and Ogorzalek, M. (eds.). 2000, 10(3).
209. *International Journal of Circuit Theory and Applications.* Special issue: Communications, Information Processing and Control Using Chaos. Hasler, M., and Vandewalle, J. (eds.). 1999, 27(6).
210. Isidori, A. *Nonlinear Control Systems*, 3rd edn. Springer-Verlag, New-York, 1995.
211. Jackson, E.A. *Perspectives of Nonlinear Dynamics, Vol. 12.* Cambridge University Press, Cambridge, England, 1990.
212. Jackson, E.A., and Grosu, I. An OPCL control of complex dynamic systems. *Physica D* 1995, 85, 1–9.
213. Janson, N.B., Balanov, A.G., and Shöll, E. Delayed feedback as a means of control of noise-induced motion. *Phys. Rev. Lett.* 2004, 93(1), 010601.
214. Jaynes, E.T. Information Theory and Statistical Mechanics, I, II. *Phys. Rev.* 1957, 106, 620–630; *Phys. Rev.* 1957, 108, 171–190.
215. Jaynes, E.T. 1988, The evolution of Carnot's principle. In: *Maximum-Entropy and Bayesian Methods in Science and Engineering.* Erickson G.J. and Smith C.R. (eds.). Kluwer, Dordrecht, pp. 267–282.
216. Judson, R.S., and Rabitz, H. Teaching Lasers to Control Molecules. *Phys. Rev. Lett.* 1992, 68, 1500–1503.
217. Just, W., Bernard, T., Ostheimer, M., Reibold, E., and Benner, H. Mechanism of time-delayed feedback control. *Phys. Rev. Lett.* 1997, 78, 203–206.
218. Just, W., Reibold, E., and Benner, H., et al. Limits of time-delayed feedback control. *Phys. Lett. A* 1999 254, 158–164.
219. Josić, K. Invariant manifolds and synchronization of coupled dynamical systems. *Phys. Rev. Lett.* 1998, 80, 3053–3056.
220. Kadomtsev, B.B. Dynamics and information. *Physics-Uspekhi*, 1994, 37(5), 425–499.
221. Kadomtsev, B.B. *Dynamics and Information.* UFN, Moscow, 1999 (in Russian).
222. Kalman, R.E., Falb, P.L., and Arbib, M.A. *Topics in Mathematical System Theory.* McGraw-Hill, New York, 1969.
223. Kapitaniak, T. *Chaos for Engineers. – Theory, Applications, and Control*, 2nd ed. Springer, Berlin-Heidelberg, 2000.
224. Kapitsa, P.L. Dynamic stability of a pendulum with oscillating suspension point. *Zh. Exper. Teor. Phys.* 1951, 21(5), (in Russian).
225. Kapur, J.N. *Maximal Entropy Models in Science and Engineering.* Wiley Eastern Ltd., New Delhi, 1989.

226. Katok, A., and Hasselblat, B. Introduction to the Modern Theory of Dynamical Systems. Encyclopedia of Mathematics and Its Applications, No. 54. Cambridge University Press, London, 1997.
227. Khalil, H.K. *Nonlinear Systems,* 3rd edn. Prentice-Hall, Upper Saddle River, 2002.
228. Khramov, A.E., and Rempen, I.S. Impact of Delayed-time Feedback on Complex Dynamics in the Hydrodynamic Pierce Model. *Radiotekh. Elektron.* 2002, 47(6), 732–738.
229. Khryashchev, S.M., and Osipenko, G.S. Studying controllability by symbolic dynamics methods. *E-journal* "Differential equations and control processes" http://www.neva.ru/journal, 1997, (1), 78–90.
230. Khryashchev, S.M. Estimation of Transport Times for Chaotic Dynamical Control Systems. In: *Proc. Intern.Conf. "Physics and Control,"* A.L. Fradkov and A.N. Churilov (eds.) *Vol. 2.* Control of Oscillations and Chaos. St. Petersburg, 2003, pp. 528–533.
231. Khryashchev, S.M. Estimation of control time for systems with chaotic behavior. *Autom. Remote Control Part I* 2004, 64(10), 1566–1579; *Part II,* 64(11), 1782–1792.
232. Kim, M., Bertram, M., Pollmann, M., Oertzen, A., Mikhailov, A.S., Rotermund, H.H., and Ertl, G. Controlling Chemical Turbulence by Global Delayed Feedback: Pattern Formation in Catalytic CO Oxidation on Pt(110). *Science* 2001, 292(18), 1357–1360.
233. Kipsnis, M.M. Chaotic phenomena in the deterministic one-dimensional pulse-width control system. *Izv. Ross. Akad. Nauk, Tekh. Kibern.* 1992, 1, 108–112 (Transl. in J. Computer and Systems Sci. International).
234. Kitaev, A.Yu., Shen, A.H., and Vyalyi, M.N. Classical and Quantum Computation. *Am. Math. Soc.* 2002 (in Russian: 1999).
235. Kittel, A., Parisi, J., and Pyragas, K. Delayed feedback control of chaos by self-adapted delay time. *Phys. Lett. A* 1995, 198, 433.
236. Klimontovich, Yu.L. *Statistical Theory of Open Systems, I.* Kluwer Academic Publisher, Dordrecht, 1995.
237. Klotz, A., and Brauer, K. A small-size neural network for computing with strange attractors. *Neural Networks* 1999, 12, 601–607.
238. Kocarev, L., Tasev, Z., and Parlitz, U. Synchronizing spatiotemporal chaos of partial differential equations. *Phys. Rev. Lett.* 1997, 79, 51–54.
239. Konishi, K., Hirai, M., Kokame, H. Sliding mode control for a class of chaotic systems. *Phys. Lett. A* 1998, 245, 511–517.
240. Konjukhov, A.P., Nagibina, O.A., and Tomchina O.P. Energy based double pendulum control in periodic and chaotic mode. In: *Proc. of 3rd Intern. Conf. on Motion and Vibration Control (MOVIC'96).* Chiba, Japan, 1996, pp. 99–104.
241. Kosloff, R., Rice, S.A., Gaspard, P., Tersigni, S., and Tannor, D.J. Wavepacket dancing: Achieving chemical selectivity by shaping light-pulses. *Chem. Phys.* 1989, 139, 201.
242. Krempl, S., Eisenhammer, T., Hubler, A., Mayer-Kress, G., and Milonni, P.W. Optimal stimulation of a conservative nonlinear oscillator: Classical and quantum-mechanical calculations. *Phys. Rev. Lett.* 1992, 69(3), 430.
243. Krotov, V.F. Global methods to improve control and optimal control of resonance interaction of light and matter. Modeling and Control of Systems.

In: *Proceedings of the Bellman Continuum Workshop 1988*. Springer-Verlag, Berlin, New York, Paris, 1988.
244. Krotov, V.F. *Global Methods in Optimal Control Theory*. Marcel Dekker, CRC Press. New-York, Basel, Hong Kong, 1996.
245. Krstić M., Kanellakopoulos, I., and Kokotović P.V. *Nonlinear and Adaptive Control Design*. Wiley, New York, 1995.
246. Kukushkin, S.A., and Osipov, A.V. Kinetics of thin film nucleation from multicomponent vapor. *J. Phys. Chem. Solids* 1995, 56(6), 831–838.
247. Kukushkin, S.A., and Osipov, A.V. Morphlogical stability of islands upon thin film condensation. *Phys. Rev. E* 1996, 53, 4964–4968.
248. Kullback, S., and Leibler, R.A. On information and sufficiency. *Ann. Math. Stat.* 1951, 22(1), 7986.
249. Kuznetsov, Yu.I., Migulin, V.V., Minakova, I.I., and Silnov, B.A. Synchronization of chaotic auto-oscillations. *Sov. Phys. Doklady* 1984, 29 318–321.
250. Kuznetsov, Yu.I., Landa, P.S., Olkhovoj, A.F., and Perminov, S.M. Relation between amplitude synchronization threshold and entropy in stochastic autooscillatory systems. *Sov. Phys. Doklady* 1985, 30, 221–224.
251. Kwon, O.J. Targeting and stabilizing chaotic trajectories in the standard map. *Phys. Lett. A* 1999, 258, 229–236.
252. Lanczos, C. *The Variational Principles of Mechanics*. Toronto University Press, Toronto, 1964.
253. Landa, P.S. *Nonlinear Oscillations and Waves in Dynamical Systems*. Kluwer Academic Publisher, Dordrecht, 1996.
254. Landa, P.S. Theory of Fluctuational Transitions and Its Applications. *J. Communications Technology and Electronics*, 2001, 46 (10), 1069–1107.
255. Landa, P.S., and Rosenblum, M.G. Synchronization and chaotization of oscillations in coupled self-oscillating systems. *Appl. Mech. Rev.* 1993, 46(7), 414–426.
256. Landau, I.D., and Lifshitz, E.M. *Statistical Physics*, Part 1. Pergamon Press, Oxford, 1980.
257. Landauer, R., Irreversibility and heat generation in the computing process. *IBM J. Res. Develop* 1961, 3, 183–191.
258. Landauer, R. Energy Requirements in Communication. *Appl. Phys. Lett.* 1987, 51, 2056–2058.
259. Landauer, R. The physical nature of information. *Phys. Lett. A* 1996, 217, 188–193.
260. Landauer, R. Minimal energy requirements in communication. *Science* 1996, 272, 1914–1918.
261. Landauer, R. Information is a physical entity. *Physica A* 1999, 263, 63–67.
262. Laser control of quantum dynamics. Special Issue: *Chem. Phys.* 2001, 267 (1–3).
263. Leff, H.S., and Rex, A.F. (eds). *Maxwell's Demon 2: Entropy, Classical and Quantum Information, Computing*, 2nd edn. Institute of Physics, Washington DC, 2003.
264. Leonov, G.A. Frequency criterion of stabilization by external harmonic action. *Avtomatika i Telemekhanika* 1986 (1), 169–174 (in Russian).
265. Leonov, G.A., Ponomarenko, D.V., and Smirnova, V.B. *Frequency Methods for Nonlinear Analysis. Theory and Applications*. World Scientific, Singapore, 1996.
266. Leonov, G.A., and Smirnova, V.B. *Mathematical Problems of Phase Synchronization Theory*. Nauka, St. Petersburg, 2000 (in Russian).

267. Li, T., and Yorke, J.A. Period three implies chaos. *Am. Math. Monthly* 1975, 82, 985–992.
268. Liao, T.L. Observer-based approach for controlling chaotic systems. *Phys. Rev. E* 1998, 57, 1604–1610.
269. Liao, T.L., and Huang, N.S. Control and synchronization of discrete-time chaotic systems via variable structure control technique. *Phys. Lett. A* 1997, 234, 262–268.
270. Lima, R., and Pettini, M. Suppression of chaos by resonant parametric perturbations. *Phys. Rev. A* 1990, 41, 726–733.
271. Lima, R., and Pettini., M. Parametric resonant control of chaos. *Int. J. Bifurcation Chaos* 1998, 8, 1675–1684.
272. Lindsey, W. *Synchronization Systems in Communications and Control.* Prentice-Hall, New Jersey, 1972.
273. Lions, J.L. On the controllability of distributed systems. *Proc. Natl. Acad. Sci. USA* 1997, 94(10) 4828–4835.
274. Liu, W.K., Wu, B., and Yuan, J.M. Nonlinear dynamics of chirped pulse excitation and dissociation of diatomic molecules. *Phys. Rev. Lett.* 1995, 75(7), 1292.
275. Lloyd, S. Quantum-mechanical Maxwell's Demon. *Phys. Rev.* 1997, A56(5), 3374–3382.
276. Lorenz, E.N. Deterministic nonperiodic flow. *J. Atmosferic Sci.* 1963, 20(2), 130–141.
277.Ločher, M., et al. Theory of controlling stochastic resonance. *Phys. Rev. E* 2000, 62(1), 317–327.
278. Loskutov, A., Rybalko, S.D., and Akinshin, L.G. Controlling dynamical systems and suppression of chaos. *Differential Eqs.* 1998, 34(8), 1143–1144.
279. Luthje, O., Wolff, S., and Pfister, G. Control of chaotic Taylor-Couette flow with time-delayed feedback. *Phys. Rev. Lett.* 2001, 86, 1745–1748.
280. Mabuchi, H., and Khaneja, N. Principles and applications of control in quantum systems. *Int. J. Robust Nonlinear Control* 15, 2005, 647–667.
281. Mackey, M.C., and Glass, L. Oscillations and chaos in physiological control systems. *Science* 1997, 197, 287–280.
282. Magnitskii, N.A, and Sidorov, S.V. Control of chaos in nonlinear dynamical systems. *Differential Eqs.* 1998, 34, 1501–1509.
283. Magnitskii, N.A., and Sidorov, S.V. Some approaches to the control problem for diffusion chaos. *Differential Eqs.* 1999, 35(5), 669–674.
284. Mareels, I.M.Y., and Bitmead, R.R. Non-linear dynamics in adaptive control: Chaotic and periodic stabilization. *Autom. I* 1986, 22, 641–655; *Autom. II*: 1988, 24, 485–497.
285. Marino, F., Barland, S., and Balle, S. Single-mode operation and transverse-mode control in VCSELs induced by frequency-selective Feedback. *IEEE Photonics Technol. Lett.* 2003, 15(6), 789.
286. Marotto, F.R. Snap-back repellers imply chaos in $\mathbb{R}^n$. *J. Math. Anal. Appl.* 1978, 63, 199–223.
287. Mathematics and computers in simulation. Special Issue on *Chaos Synchronization and Control.* E. Mosekilde, A.L. Fradkov, and I.I. Blekhman (eds.) 2002, 58(4–6).
288. Matsumoto, K., and Tsyda, I. Noise induced order. *J. Stat. Phys.* 1983, 31, 87–106.

289. *Maximum-Entropy and Bayesian Methods in Science and Engineering.* G.J. Erickson and C.R. Smith (eds.), Kluwer, Dordrecht, 1988.
290. Maza, D., Mancini, H., Boccaletti, S., Genesio, R., and Arecchi, FT. Control of amplitude turbulence in delayed dynamical systems. *Int. J. Bifurcation and Chaos* 1998, 8, 1843–1848.
291. McGeer, T. Dynamics and control of bipedal locomotion. *J. Theor. Biol.* 1993, 163, 277–314.
292. Meerkov, S.M. Principle of vibrational control: Theory and applications. *IEEE Trans. Autom. Control* 1980, AC-25, 755–762.
293. Meerov, M.V. On the Routh-Hurwitz Problem for Equations with a Small Parameter. *Autom. Remote Control* 1991, 52 (7).
294. Meerson, B., and Friedland, L. Strong autoresonance excitation of Rydberg atoms: The Rydberg accelerator. *Phys. Rev. A* 1990, 41, 5233.
295. Meucci, R., Labate, A., and Ciofini, M. Controlling chaos by negative feedback of subharmonic components. *Phys. Rev. E* 1997, 56, 2829–2834.
296. Minty, M.G., and Zimmermann, F. *Measurement and Control of Charged Particle Beams.* Series: Particle Acceleration and Detection. Springer, Berlin Heidelberg 2003.
297. Mironova, M.A., Amelkin, S.A., and Tsirlin, A.M. *Mathematical Methods of Finite-time Thermodynamics.* Khimia, Moscow, 2000 (in Russian).
298. Miroshnik, I.V., and Ushakov, A.V. Design of the algorithm of synchronous control of the quasisimilar systems. *Autom. Remote Control* 1977, 38 (11).
299. Miroshnik, I.V. *Coordinated Control of Multichannel Systems.* Energoatomizdat, Leningrad, 1990 (in Russian).
300. Mirrahimi, M., and Rouchon, P. Trajectory generation for quantum systems based on Lyapounov techniques. In: *Prepr. 6th IFAC Symposium on Nonlinear Control Systems (NOLCOS 2004)*, Stuttgart, Germany, 2004, pp. 311–316.
301. Mirrahimi, M., Turinici, G., and Rouchon, P. Reference trajectory tracking for locally designed coherent quantum controls. *J. Phys. Chem. A* 2005, 109, 2631–2637.
302. Mirzoev, F.Kh., Panchenko, V.Ya., and Shelepin L.A. Laser control processes in solids. Phys. Uspekhi 1996, 39(1), 1–29.
303. Montagne, R., and Colet, P. Nonlinear diffusion control of spatiotemporal chaos in the complex Ginzburg-Landau equation. *Phys. Rev. E* 1997, 56, 4017–4024.
304. Moon, F. *Chaotic and Fractal Dynamics. An Introduction for Applied Scientists and Engineers.* Wiley, Hoboken, NJ, 1992.
305. Morgul, O. On the control of chaotic systems in Lur'e form by using dither. *IEEE Trans. Circ. Syst. I* 1999, 46, 1301–1305.
306. Mormann, F., Lehnertz, K., David, P., and Elger, C.E. Mean phase coherence as a measure for phase synchronization and its application to the EEG of epilepsy patients. *Physica D* 2000, 144(3–4), 358–369.
307. Mosekilde, E., and Mosekilde, L., (eds). *Complexity, Chaos, and Biological Evolution, NATO ASI Series, Series B: Physics, Vol. 270*, Plenum Press, New York London, 1991.
308. Morgul, O. On the control of some chaotic systems by using dither. *Phys. Lett. A* 1999, 262, 144–151.
309. Nakagawa, M. A chaos associative memory model with a sinusoidal activation function. *Chaos Solitons Fractals* 1999, 10, 1437–1452.
310. Nakajima, H. On analytical properties of delayed feedback control of chaos. *Phys. Lett. A* 1997, 232, 207–210.

311. Nakar, E., and Friedland, L. Passage through resonance and autoresonance in x^{2n}-type potentials. *Phys. Rev. E* 1999, 60(5), 5479–5485.
312. Neimark, Yu.I., and Landa, P.S. *Stochastic and Chaotic Oscillations.* Kluwer Academic Press, Dordrecht, 1992 (translated from the Russian edition, 1987).
313. von Neumann, J. *Theory of Self-Reproducing Automata*, edited and completed by Arthur W. Burks, University Illinois Press, Urbana and London, 1966.
314. Nielsen, M.A., and Chuang, I.L. *Quantum Computation and Quantum Information.* Cambridge University Press, Cambridge, 2000.
315. Nijmeijer, H., and Mareels, I.M.Y. An observer looks at synchronization. *IEEE Trans. Circ. Syst. I* 1997, 44, 882–890.
316. Nijmeijer H., and van der Schaft, A.J. *Nonlinear Dynamical Control Systems.* Springer-Verlag, New York, 1990.
317. Nijmeijer, H., and Rodriguez-Angeles, A. *Synchronization of Mechanical Systems.* World Scientific, Singapore, 2003.
318. Nikolis, G., and Prigogine, I. *Selforganization in Nonequilibrium Systems.* Wiley, New York, 1977.
319. Norton, J.D. Eaters of the lotus: Landauer's principle and the return of maxwell's demon. *Stud. Hist. Philos. Mod. Phys.* 2005, 36, 375–411.
320. Novikov, I.I. The efficiency of atomic power stations. *Atomic Energy* 3(11), 409–412, 1957 (English translation: *Nuclear Energy II* 7, 125–128, 1958).
321. Nyquist, H. Certain factors affecting telegraph speed. *Bell System Technical J.* 1924, 3, 324–346.
322. Ogorzalek, M.J. Taming chaos – part I: Synchronization; part II: Control. *IEEE Trans. Circ. Syst. Part I* 1993, 40, 639–706.
323. Onsager, L. Reciprocal relations in irreversible processes. *Phys. Rev.* 1931, 37, 405–426.
324. Ortega, R., Loria A., Nicklasson, P.J., and Sira-Ramirez, H. *Passivity-based Control of Euler–Lagrange Systems.* Springer, Berlin, 1998.
325. Ortega, R., van der Schaft, A., Mareels, I., and Maschke, B. Putting energy back in control. *IEEE Control Systems Magazine* 2001, 21(2), 18–33.
326. Ortega, R., van der Schaft, A.J., Maschke, B., and Escobar, G. Interconnection and damping assignment passivity-based control of port-controlled Hamiltonian systems. *Automatica* 2002, 38 585–596.
327. Osipenko, G.S., and Khryashchev, S.M. Controllability and applied symbolic dynamics. In: *Proc.13th Intern. Symp. on Mathematical Theory of Networks and Systems (MTNS'98)*. Padova, Italy, 1998, pp. 349–352.
328. Osipov, G.V., Pikovsky, A., Rosenblum, M., and Kurths, J. Phase synchronization effects in a lattice of nonidentical Rössler oscillators. *Phys. Rev. E* 1997, 55(3), 2353–2361.
329. Osipov, G.V., and Kurths, J. Regular and chaotic phase synchronization of coupled circle maps. *Phys. Rev. E* 65, 2002, 016216.
330. Osipov, G.V., Bambi, Hu, Zhou, C., Ivanchenko, M.V., and Kurths, J. Three types of transitions to phase synchronization in coupled chaotic oscillators. *Phys. Rev. Lett.* 2003, 91(2), 4101–4104.
331. Ott, E., Grebogi, C., and Yorke, J. Controlling chaos. *Phys. Rev. Lett.* 1990, 64(11), 1196–1199.
332. Ott, E. *Chaos in Dynamical Systems.* 2nd Edition. New York: Cambridge University Press, 2002.
333. Ohtsuki, Y. *Chem. Phys. Lett.* 2005, 404, 126.

334. Owens, D.H. Robust stability of Smith predictor controllers for time-delay systems. *Proc. IEE Part D* 1982, 129, 298–304.
335. Park, K.S., Park, J.B., Choi, Y.H., et al. Generalized predictive control of discrete-time chaotic systems. *Int. J. Bifurcation Chaos*. 1998, 8, 1591–1597.
336. Park, J., Kim, J., Cho, S.H., Han, K.H., et al. Development of sorbent manufacturing technology by agitation fluidized bed granulator. *Korean J. Chem. Eng.* 1999, 16(5), 659–663.
337. Park, M.S., Ahn, B.T., Yoo, B.-S., Chu, H.Y., Park, H.-H., and Chang- Hasnain, C.J. Polarization control of vertical-cavity surface-emitting lasers by electro-optic birefringence. *Appl. Phys. Lett.* 2000, 76, 813–815.
338. Parmananda, P., Hildebrand, M., and Eiswirth, M. Controlling turbulence in coupled map lattice systems using feedback techniques. *Phys. Rev. E* 1997, 56, 239–244.
339. Paskota, M., and Lee, H.W.J. Targeting moving targets in chaotic dynamical systems. *Chaos, Solitons Fractals* 1997, 8, 1533–1544.
340. Paskota, M., Mees, A.I., and Teo, K.L. Directing orbits of chaotic systems in the presence of noise: Feedback correction. *Dynamics Control* 1997, 7, 25–47.
341. Pavon, M., and Ticozzi, F. Controlling the density evolution of classical, thermodynamic and quantum systems. In: *Proc. 44th IEEE Conf. Dec. Control.* 2005, pp. 1800–1805.
342. Pearson, B.J., White, J.L., Weinacht, T.C., and Bucksbaum., P.H. Coherent control using adaptive learning algorithms. *Phys. Rev. A* 2001, 63, 063412.
343. Pecora, L.M., and Carroll., T.L. Synchronization in chaotic systems. *Phys. Rev. Lett.* 1990, 64, 821–823.
344. Peirce, A., Dahleh, M., and Rabitz, H. Optimal control of quantum mechanical systems: Existence, numerical approximations, and applications. *Phys. Rev. A* 1988, 37, 4950.
345. Petrov, B.N., Ulanov, G.M., Goldenblatt, I.I., and Ulyanov, S.V. *Control Problems for Relativistic and Quantum Dynamical Systems*. Nauka, Moscow, 1982.
346. Petrov, V., Gaspar, V., Masere, J., and Showalter, K. Controlling chaos in the Belousov-Zhabotinsky reaction. *Nature* 1993, 361, 240–243.
347. Pettini, M. Controlling chaos through parametric excitations In: Dynamics and Stochastic Processes, R. Lima, L. Streit, and R.V. Vilela-Mendes (eds.). Springer-Verlag, New York, 1988, pp. 242–250.
348. Piel, A., Greiner, F., Klinger, T., Krahnstover, N., and Mausbach, T. Chaos and chaos control in plasmas. *Physica Scripta* 2000, T84, 128–131.
349. Pikovsky, A., Rozenblum, M., Osipov, G., and Kurths, J. Phase synchronization of chaotic oscillators by external driving. *Physica D* 1997, 104, 219–238.
350. Pikovsky, A., Rosenblum, M., and Kurths, J. *Synchronization. A Universal Concept in Nonlinear Sciences*. Cambridge University Press, Cambridge, 2001.
351. Place, C.M., and Arrowsmith, D.K. Control of transient chaos in tent maps near crisis. *Phys. Rev. E* 2000, 61, I: 1357–1368; II: 1369–1381.
352. Pogromsky, A., Santoboni, G., and Nijmeijer, H. Partial synchronization: From symmetry towards stability. *Physica D* 2002, 172, 65–87.
353. Plank, M. Das Prinzip der kleinsten Wirkung. *Die Kultur der Gegenwart*, 1914, 3, Abt. 3, Bd. 1, ss. 692–702 (Also in: Physicalishe Abhandlungen und Vortrage, Bd.3, Braunshweig, 1958, ss. 91–101).
354. Polushin, I.G., L.Fradkov, A., and Hill, D.J. Passivity and passification of nonlinear systems. *Autom. Remote Control* 2000, 61(3, Part 1), 355–388.

355. Poplavskii, R.P. Maxwell demon and the correspondence between information and entropy. *Phys. Uspekhi* 1979, 22(5), 371–380.
356. Poplavskii, R.P. *Thermodynamics of Information Processes*. Nauka, Moscow, 1981.
357. Postnikov, N.S. Stochasticity of the relay hysteresis systems. *Autom. Remote Control* 1998, 59, 349–358.
358. *Proc. Intern. Conf. "Physics and Control."* A.L. Fradkov and A.N. Churilov (eds.). *Vol. 1: Physics and Control: General Problems and Applications.* IEEE, St. Petersburg, Aug. 2003.
359. *Proc. Intern. Conf. "Physics and Control."* A.L. Fradkov and A.N. Churilov (eds.). *Vol. 2: Control of Oscillations and Chaos.* IEEE, St. Petersburg, Aug. 2003.
360. *Proc. Intern. Conf. "Physics and Control."* A.L. Fradkov and A.N. Churilov (eds.). *Vol. 3: Control of Microworld Processes. Nano- and Femtotechnologies.* IEEE, St. Petersburg, Aug. 2003.
361. *Proc. Intern.Conf. "Physics and Control."* A.L. Fradkov and A.N. Churilov *Vol. 4: Nonlinear Dynamics and Control.* IEEE, St. Petersburg, Aug. 2003.
362. Pyragas, K. Continuous control of chaos by self-controlling feedback. *Phys. Lett. A* 1992, 170, 421–428.
363. Pyragas, K. Control of chaos via an unstable delayed feedback controler. *Phys. Rev. Lett.* 2001, 86, 2265–2268.
364. Rabinovich, M.I., Abarbanel, H.D.I., Huerta, R., Elson, R., and Selverston, A.I. Self-regularization of chaos in neural systems: Experimental and theoretical results. *IEEE Trans. Circuits Syst. I* 1997, 44, 997–1005.
365. Rabinovich, M.I., Varona, P., and Abarbanel, H.D.I. Nonlinear cooperative dynamics of living neurons. *Int. J. Bifurcation Chaos* 2000, 10, 913–933.
366. Rabitz, H. Algorithms for closed loop control of quantum dynamics. In: *Proc. 39th IEEE Conf. Decisions and Control.* Sydney, 12–15 Dec. 2000, 937–942.
367. Rabitz, H. Controlling quantum phenomena: The dream is becoming a reality In: *Preprints of the 5th IFAC Symp. Nonlin. Contr. Systems (NOLCOS'01).* St. Petersburg, 4–6 July, 2001.
368. Rabitz, H., Turinici, G., and Brown, E. Control of molecular motion: Concepts, procedures, and future prospects. In: *Handbook of Numerical Analysis*, Vol. X. P. Ciarlet and J. Lions (eds.). Elsevier, Amsterdam, 2003.
369. Ramaswamy, R., Sinha, S., and Gupte, N. Targeting chaos through adaptive control. *Phys. Rev. E* 1998, 57, R2507–R2510.
370. Rangan, C., Bloch, A.M., Monroe, C., and Bucksbaum, P.H. Control of trapped ion quantum states with optical pulses. *Phys. Rev. Lett.* 2004, 92, 113004.
371. Ravindra, B., and Hagedorn, P. Invariants of chaotic attractor in a nonlinearly damped system. *J. Appl. Mech. Trans. ASME* 1998, 65, 875–879.
372. Rice, J. Active control of molecular dynamics: Coherence versus chaos. *J. Stat. Phys.* 2000, 101, 187.
373. Rosenblueth, A., Wiener, N., and Bigelow, J. Behavior, purpose and teleology. *Philosophy Sci. Baltimore* 1943, 10(1), 18–24.
374. Rosenblum, M.G., Firsov, G.I., Kuuz, R.A., and Pompe, B. Human postural control: Force plate experiments and modelling. In: Nonlinear Analysis of Physiological Data. H. Kantz, J. Kurths, and G. Mayer-Kress (eds.). Springer-Verlag, Berlin, 1998, 283–306.
375. Rosenblum, M.G., Pikovsky, A.S., and Kurths, J. Phase synchronization of chaotic oscillators. *Phys. Rev. Lett.* 1996, 76, 1804–1807.

376. Rosenblum, M.G., Pikovsky, A.S., and Kurths, J. From phase to lag synchronization in coupled chaotic oscillators. *Phys. Rev. Lett.* 1997, 78, 4193–4196.
377. Rosenbrock, H.H. A stochastic variational principle for quantum mechanics. *Phys. Lett.* 1986, 110A, 343–346.
378. Rosenbrock, H.H. *Machines with a Purpose*. Oxford University Press, New York, 1990.
379. Rosenbrock, H.H. Doing quantum mechanics with control theory. *IEEE Trans. Autom. Control* 2000, 45(1), 73–77.
380. Rouche, N., Habets, P., and Laloy, M. *Stability Theory by Lyapunov's Direct Method*. Springer-Verlag, New York, 1977.
381. Roy, R., Murphy, T.W., Maier, T.D., Gills, Z., and Hunt, E.R. Dynamical control of a chaotic laser: Experimental stabilization of a globally coupled system. *Phys. Rev. Lett.* 1992, 68, 1259–1262.
382. Roy Chowdhury, A., Saha, P., and Banerjee, S. Control of chaos in laser plasma interaction. *Chaos Solitons Fractals* 2001, 12, 2421–2426.
383. Ruelle, D., and Takens, F. On the nature of turbulence. *Comm. Math. Phys.* 1971, 20(2), 167–192.
384. Rulkov, N.F., Sushchik, M., Tsimring, L.S., and Abarbanel, H.D.I. Generalized synchronization of chaos. *Phys. Rev. E* 1995, 51, 980.
385. Sagdeev, R.Z., Usikov, D.A., and Zaslavsky, G.M. *Nonlinear Physics: From Pendulum to Turbulence and Chaos*. Harwood Academic Publisher, Chur, 1988.
386. Sakurai, T., Mihaliuk, E., Chirila, F., and Showalter, K. Design and control of wave propagation patterns in excitable media. *Science* 2002, 296(14), 2009–2012.
387. Salamon, P., Nulton, J.D., Siragusa, G., Andersen, T.R., and Limon, A. Principles of control thermodynamics. *Energy* 2001, 26, 307–319.
388. Sastry, S.S., and Bodson, M. *Adaptive Control: Stability, Convergence and Robustness*. Prentice-Hall, Englewood Cliffs, NJ, 1989.
389. Schäfer, C., Rosenblum, M.G., Kurths, J., and Abel, H.-H. Heartbeat synchronized with ventilation. *Nature* 1998, 392(6673), 239–240.
390. Van der Schaft, A.J. L^2-*gain and Passivity Techniques in Nonlinear Control*, 2nd ed. Springer, Berlin, 1999.
391. Schiff, S.J., Jerger, K., Duong, D.H., Chang, T., Spano, M.L., and Ditto, W.L. Controlling chaos in the brain. *Nature* Aug. 25, 1994, 370, 615–620.
392. Schomburg, E., Hofbeck, K., Scheuerer, R., et al. Control of the dipole domain propagation in a GaAs/AlAs superlattice with a high-frequency field. *Phys. Rev. B* 2002, 65, 155320.
393. Schuster, H.G., and Stemmler, M.B. Control of chaos by oscillating feedback. *Phys. Rev. E* 1997, 56, 6410–6417.
394. Schwartz, I.B., and Triandaf, I. Tracking sustained chaos. *Int. J. Bifurcation Chaos* 10, 2000, 571–578.
395. Schwieters, C.D., and Rabitz, H. Optimal control of nonlinear classical systems with application to unimolecular dissociation reactions and chaotic potential. *Phys. Rev. A* 1991, 44(8), 5224.
396. *Self-Organized Biological Dynamics and Nonlinear Control. Toward Understanding Complexity, Chaos and Emergent Function in Living Systems*. J. Walleczek (ed.). Cambridge University Press, Cambridge, 2000.
397. Sen, A.K. Control and diagnostic uses of feedback. *Phys. Plasmas* 2000, 7 1759–1766.

398. Shapiro, M., and Brumer, P. Laser control of product quantum state populations in unimolecular reactions. *J. Chem. Phys.* 1986, 84, 4103–4110.
399. Sharma, A., and Gupte, N. Control methods for problems of mixing and coherence in chaotic maps and flows. *Pramana J. Phys.* 1997, 48, 231–248.
400. Sharkovsky, A.N. Coexistence of cycles of a continuous map of the line into itself. *Ukrainian Math. J.* 1964, 1, 61–71 (Translated in: *Int. J. Bifurcation Chaos* 5(5), 1263–1273, 1995).
401. Shannon, C. A Mathematical theory of communication. *Bell. Syst. Tech. J.* 27, 1948, 3, 379–423; 4, 623–656.
402. Shilnikov, L.P. The existence of a denumerable set of periodic motions. *Sov. Math. Dokl.* 6, 1965, 163–166.
403. Shinbrot, T., Grebogi, C., Ott, E., and Yorke, J.A. Using small perturbations to control chaos. *Nature* 1993, 363, 411–417.
404. Shiriaev, A.S., and Fradkov, A.L. Stabilization of invariant sets for nonaffine nonlinear systems. *Automatica* 2000, 36(11), 1709–1715.
405. Seeger, R.J. *Galileo Galilei, His Life and His Works*, Pergamon, Oxford, 1966.
406. *Selected Topics in Vibrational Mechanics*. I.I. Blekhman (ed.). World Scientific, Singapore, 2004.
407. Shiriaev, A.S., and Fradkov, A.L. Stabilization of invariant sets for nonlinear systems with application to control of oscillations. *Intern. J. Robust Nonlinear Control* 2001, 11, 215–240.
408. Shiriaev, A.S., Egeland, O., Ludvigsen, H., and Fradkov, A.L. VSS-version of energy-based control for swinging up a pendulum. *Syst. Control Lett.* 2001, 44(1), 45–56.
409. Schöll, E. Nonlinear Spatio-Temporal Dynamics and Chaos in Semiconductors. Cambridge University Press, Cambridge, 2001.
410. Sieniutycz, S., and Farkas, H. (eds.). *Variational and Extremum Principles in Macroscopic Systems*. Elsevier Science, Oxford, 2004.
411. Singer, J., Wang, Y.Z., and Bau, H.H. Controlling a chaotic system. *Phys. Rev. Lett.* 1991, 66, 1123–1125.
412. Sinha, S., and Gupte, N. Adaptive control of spatially extended systems: Targeting spatiotemporal patterns and chaos. *Phys. Rev. E* 1998, 58(5), R5221–R5224.
413. Smith, O.J.M. Closer control of loops with dead time. *Chem. Eng. Progress* 1957, 53, 217–219.
414. Sontag, E.D. *Mathematical Control Theory, Deterministic Finite Dimensional Systems*, 2nd edn. Springer-Verlag, New York, 1998.
415. Sontag, E. Structure and stability of certain chemical networks and applications to the kinetic proofreading model of T-cell receptor signal transduction. *IEEE Trans. Autom. Control* 2001, 46, 1028–1047.
416. Sontag, E., Kholodenko, B.N., Kiyatkin, A., Bruggeman, F., Westerhoff, H., and Hoek, J. Untangling the wires: A novel strategy to trace functional interactions in signaling and gene networks. *Proc. Natl. Acad. Sci. USA* 2002, 99, 12841–12846.
417. Stephenson, A. On a new type of dynamical stability. *Mem. Proc. Manch. Lit. Phil. Soc.* 1908, 52, 1–10; On induced stability. *Phil. Mag.* 1909, 15, 233–236.
418. Stewart, H.B., Thompson, J.M.T., Ueda, U., and Lansbury, A.N. Optimal escape from potential wells – patterns of regular and chaotic bifurcations. *Physica D* 1995, 85, 259–295.

419. Strogatz, S. *Nonlinear Dynamics and Chaos*. Addison-Wesley, Reading, 1994.
420. Suzuki, S., Mishima, K., and Yamashita, K. Ab initio study of optimal control of ammonia molecular vibrational wavepackets: Towards molecular quantum computing. *Chem. Phys. Lett.* 2005, 410, 358–364.
421. Svirezhev, Yu.M., and Logofet, D.O. *Stability of Biological Communities*. Mir, Moscow, 1983 (in Russian: 1978).
422. *Systems and Control Letters*. Special issue: Control and synchronization of chaos. H. Nijmeijer (ed.). 1997, 34(5).
423. Szillard, L. On the decrease of entropy in a thermodynamic system by the intervention of intelligent beings. *Z. Phys.* 1929, 53, 840.
424. Tanaka, T., Inoue, M., and Fujisaka, H. Self-organization in a spin model of chaos neural network. *Progress Theor. Phys. Suppl.* 2000, 598–599.
425. Tannor, D.J., and Rice, S.A. Control of selectivity of chemical reaction via control of wave packet evolution. *J. Chem. Phys.* 1985, 83, 5013.
426. Tesch, C.M., de Vivie-Riedle, R. *Phys. Rev. Lett.* 89, 2002, 157901.
427. Touchette, H., and Lloyd S. Information-theoretic approach to the study of control systems. *Physica A* 2004, 331(1–2), 140–172.
428. Turchin, V.F. *The Phenomenon of Science*. Columbia University Press, New York, 1977.
429. Ullal, G.R., Dasgupta, C., and Biswal, B. Chaos control and neural network – modelling of epilepsy. *Epilepsia* 1999, 40, 208–208.
430. Umeda, H., and Fujimura, Y. Applicability of a classical local control method to a quantum system. *Chem. Phys.* 2001, 274, 231–241.
431. Ushio, T. Limitation of delayed feedback control in nonlinear discrete-time systems. *IEEE Trans. Circ. Sys. I* 1996, 43, 815–816.
432. Valle, R. *History of Cybernetics. Encyclopedia of Life Support Systems (EOLSS)*. Eolss Publishers, Oxford, UK, 2003 [http://www.eolss.net]. Developed under the auspices of the UNESCO.
433. Van der Pol, B. Forced oscillations in a circuit with nonlinear resistance. *Philos. Mag. Ser. 7* 1927, 3(13), 65–80.
434. Vanecek, A., and Celikovsky, S. Chaos synthesis via root locus. *IEEE Trans. Circ. Syst. I* 1994, 41, 59–60.
435. Vincent, T.L. Control using chaos. *IEEE Control Syst. Mag.* 1997, 17, 65–76.
436. Virgin, L.N., and Cartee, L.A. A note on the escape from a potential well. *Int. J. Nonlinear Mech.* 1991, 26, 449–458.
437. Vorotnikov, V.I. *Partial Stability and Control*. Birkhäuser, Basel, 1998.
438. Voss, H.U. Anticipating chaotic synchronization. *Phys. Rev. E* 2000, 61(5), 5115–5119.
439. Vstovsky, G.V. Interpretation of the extreme physical information principle in terms of shift information. *Phys. Rev. E* 51(2), 1995, 975–979.
440. Walker, R., and Preston, R. *J. Chem. Phys.* 1977, 67, 2017.
441. Wallraff, A., Schuster, D.I., Blais, A., et al. Strong coupling of a single photon to a superconducting qubit using circuit quantum electrodynamics. *Nature* 2004, 431, 162–167.
442. Wang, H.O., and Abed E.H. Bifurcation control of a chaotic system. *Automatica* 1995, 31 1213–1226.
443. Wang, P.Y., and Xie, P. Eliminating spatiotemporal chaos and spiral waves by weak spatial perturbations. *Phys. Rev. E* 2000, 61, 5120–5123.
444. Wang, X.F., and Chen, G.R. On feedback anticontrol of discrete chaos. *Int. J. Bifurcation Chaos* 1999, 9, 1435–1441.

445. Wheeler, J.A. In: *Complexity, Entropy and the Physics of Information*. W.H. Zurek (ed.). Addison-Wesley, New York, 1990, p. 3.
446. Wiener, N. *Cybernetics or Control and Communication in the Animal and the Machine*. MIT Press, Cambridge, MA, 1948.
447. Wiener, R.J., Dolby, D.C., Gibbs, G.C., Squires, B., Olsen, T., and Smiley, A.M. Control of chaotic pattern dynamics in Taylor vortex flow. *Phys. Rev. Lett.* 1999, 83, 2340–2343.
448. Willems, J.C. Dissipative dynamical systems. Part I: General theory. *Arch. Rational Mech. Analysis* 1972, 45(5), 321–351.
449. Wu, N. *The Maximum Entropy Method*. Springer-Verlag, New York, 1997.
450. Xiao, J.H., Hu, G., and Gao, J.H. Turbulence control and synchronization and controllable pattern formation. *Int. J. Bifurcation Chaos* 2000, 10(3), 655–660.
451. Yakubovich, V.A., Leonov, G.A., and Gelig A.Kh. *Stability of Stationary Sets in Control Systems with Discontinuous Nonlinearities Singapore*. World Scientific, Singapore, 2004.
452. Yang, L., Liu, Z., and Zheng, Y. "Middle" periodic orbit and its application to chaos control. *Intern. J. Bifurcation Chaos* 2001, 12(8), 1869–1876.
453. Yau, H.T., Chen, C.K., and Chen, C.L. Sliding mode control of chaotic systems with uncertainties. *Int. J. Bifurcation Chaos* 2000, 10, 1139–1147.
454. Ydstie, B.E. Passivity based control via the second law. *Comput. Chem. Eng.* 2002, 26(7–8), 1037–1048.
455. Yu, C., Gross, P., Ramakrishna, V., Rabitz, H., Mease, K., and Singh, H. Control of classical regime molecular objectives – applications of tracking and variations of the theme. *Automatica* 1997, 9 1617–1633.
456. Yu, X.H. Controlling chaos using input-output linearization approach. *Int. J. Bifurcation Chaos* 1997, 7, 1659–1664.
457. Yu, X.H. Variable structure control approach for controlling chaos. *Chaos, Solitons Fractals* 1997, 8, 1577–1586.
458. Vasiliev, V., Romanovsky, Yu., and Yahno, V. *Autowave Processes*. Nauka, Moscow, 1987 (in Russian).
459. Zames, G., Shneydor N.A. Dither in nonlinear systems. *IEEE Trans. Autom. Control* 1976, 21, 660–667.
460. Zewail, A. Femtochemistry: Atomic-Scale dynamics of the chemical bond. (Adapted from the Nobel Lecture). *J. Phys. Chem. A* 2000, 104, 5660–5694.
461. Zhao, H., Wang, Y.H., and Zhang, Z.B. Extended pole placement technique and its applications for targeting unstable periodic orbit. *Phys. Rev. E* 1998, 57, 5358–5365.
462. Zhou, C., and Kurths, J. Noise-induced phase synchronization and synchronization transitions in chaotic oscillators. *Phys. Rev. Lett.* 2002, 88(23), 230–602.
463. Zhou, C., Kurths, J. Kiss, I.Z., and Hudson, J.L. Noise-enhanced phase synchronization of chaotic oscillators. *Phys. Rev. Lett.* 2002, 89(1), 014–101.
464. Zhou, K., Doyle, J.C., and Glover, K. *Robust and Optimal Control*. Prentice Hall, New Jersey, 1996.

Index

adaptive observer 92, 98
Aero model 149
Anosov lemma 108
anticontrol of chaos 112
approximate synchronization 69
Astrov model 204
asymptotic synchronization 24
attracting set 107
attractor 107
autoresonance 60

Bohl exponent 109
Brunovsky form 33

Carnot's Principle 7
chaos 2, 25, 105, 106, 128, 155
chaotic attractor 107
chaotization 112
cluster synchronization 156
control algorithms 26
control goal 21, 28, 29, 181
control of chaos 2, 105
control parameter 11
control thermodynamics 9
control variable 11
controlled dissociation 164
controlled Poincaré map 111, 116
controlled Schrödinger equation 167, 168
controlled synchronization 67, 112
convergence 83
convergent system 79
coordinate synchronization 72
coordinated control 81

cross-entropy 9
cybernetical physics 3, 17, 27, 213

delayed coordinates 110, 117
delayed feedback 120
Demidovich condition 79
direct adaptive control 126
dissipation rate 52

entropy 7, 176
ε-synchronization 69
excitability index 50
external synchronization 80

feedback control 13
feedback linearization 32
feedback resonance 57
feedforward control 113
feedforward control 27
femtochemistry 162
femtotechnologies 162
Floquet matrix 120
forced synchronization 67
Frenkel-Kontorova model 140
frequency synchronization 67, 70

generalized synchronization 72
global feedback 155

high-gain observer 84, 86
Hurwitz matrix (polynomial) 80
Huygens synchronization 70, 74
hyper minimum phase 90, 91, 96
hyper-minimum-phaseness 90

Index

identification-based adaptive control 126
indirect adaptive control 126
information physics 6

Kalman controllability criterion 81
Kapitsa pendulum 183
Kukushkin-Osipov model 193
Kullback-Leibler divergence 9

lag synchronization 72
Lipschitz condition 84
Lotka-Volterra model 199
Lurie system 120
Lyapunov equation 204
Lyapunov exponent 25, 49, 109, 120, 148
Lyapunov function 32, 85, 91, 92, 127, 157, 179
Lyapunov inequality 91
Lyapunov instability 107
Lyapunov stability 107

Marotto theorem 110
master-slave synchronization 80
Maximum Entropy Principle (MEP) 7, 176
Meerov condition 80, 81
Melnikov method 115
minimum phase 90, 91
monodromy matrix 120

nonfeedback control 105, 113, 136, 153, 189
normalized excitability index 50

observability 82
observer 29, 82, 86, 121, 123, 187
occasional proportional feedback (OPF) 117
oddness limitation 120
OGY method 116
OGY-law 35
Onsager principle 180
open loop control 27
output feedback 27

partial synchronization 72
passification 87, 91
passivity 52

Pecora-Carroll synchronization scheme 84
persistent excitation 90
phase synchronization 71
Poincaré map 110, 116, 130, 192
Poincare section 71, 111, 116, 117, 121
Poisson bracket 39
precomparison functions 77
predator-prey system 199
predissociation 164, 166
principle of Minimum Fisher Information (MFI) 9
Pugh lemma 108
Pyragas method 120

quantum Poisson bracket 169

reaction–diffusion equations 137
recurrence 108
recursive goal inequalities 29, 118
reference model 23
reference signal 22
regulation 21
relative degree 81
relative entropy 9
relay speed-gradient algorithm 31
robustification 93
robustness 82
roughness 82

self-synchronization 67
SG-principle 174
Sharkovsky-Li-Yorke criterion of chaos 109
Shiriaev's condition 45, 57
small control 13, 43
speed-gradient (SG) algorithm 30
speed-gradient method 29, 123
speed-gradient principle 174
speed-pseudogradient algorithm 31
state feedback 27
Stephenson-Kapitsa pendulum 183
storage function 52
strange attractor 107
swingability 43
swinging control 43
synchronization 24, 67, 112
synchronization index 69
synchronous frequency 70

tracking 22
transformation laws 34

variational principles 173

vibrational control 12, 183
vibrational mechanics 12

zero-state detectablity 45

Understanding Complex Systems

Edited by J.A. Scott Kelso

Jirsa, V.K.; Kelso, J.A.S. (Eds.)
Coordination Dynamics: Issues and Trends
XIV, 272 p. 2004 [3-540-20323-0]

Kerner, B.S.
The Physics of Traffic:
Empirical Freeway Pattern Features,
Engineering Applications, and Theory
XXIII, 682 p. 2004 [3-540-20716-3]

Kleidon, A.; Lorenz, R.D. (Eds.)
Non-equilibrium Thermodynamics
and the Production of Entropy
XIX, 260 p. 2005 [3-540-22495-5]

Kocarev, L.; Vattay, G. (Eds.)
Complex Dynamics in Communication Networks
X, 361 p. 2005 [3-540-24305-4]

McDaniel, R.R.Jr.; Driebe, D.J. (Eds.)
Uncertainty and Surprise in Complex Systems:
Questions on Working with the Unexpected
X, 200 p. 2005 [3-540-23773-9]

Ausloos, M.; Dirickx, M. (Eds.)
The Logistic Map and the Route to Chaos:
From The Beginnings to Modern Applications
XVI, 411 p. 2006 [3-540-28366-8]

Kaneko, K.
Life: An Introduction to Complex Systems Biology
XIV, 371 p. 2006 [3-540-32666-9]

Braha, D.; Minai, A.A.; Bar-Yam Y. (Eds.)
Complex Engineered Systems: Science Meets Technology
VIII, 394 p. 2006 [3-540-32831-9]

Fradkov, A.L.
Cybernetical Physics: From Control of Chaos to Quantum Control
XII, 242 p. 2007 [3-540-46275-9]

Printing: Krips bv, Meppel
Binding: Stürtz, Würzburg